GEOLOGICAL FRAGMENTS.

GEOLOGICAL FRAGMENTS
OF
FURNESS AND CARTMEL

BY

JOHN BOLTON

With a new gazetteer
prepared expressly for this edition

By Harry Kellett

Michael Moon at the Beckermet Bookshop, Beckermet,
Cumbria and also at Michael Moon's Bookshop,
41-42 Roper Street, Whitehaven, Cumbria.

1978

First Published in 1869.
Republished with additions 1978, by Michael Moon, in a short print run of 600 copies.

ISBN. 09-04131-20-3

The publisher would like to record his thanks to Harry Kellett for his work in providing this work with a much needed Gazetteer, Contents List and Dramatis Personae.

Printed in Great Britain by
The Scolar Press, Ilkley, Yorkshire.

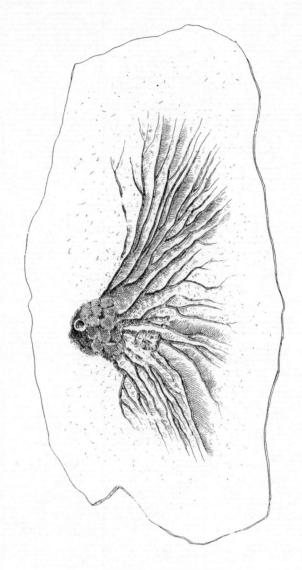

ACTINOCRINUS

(new species)

From Knott-hollow Quarry,
near Ulverston.

GEOLOGICAL FRAGMENTS

COLLECTED PRINCIPALLY

FROM

RAMBLES AMONG THE ROCKS

OF

FURNESS AND CARTMEL.

BY

JOHN BOLTON.

"The man who consecrates his hours
By vig'rous effort and an honest aim,
At once he draws the sting of life and death;
He walks with nature, and her paths are peace."—YOUNG.

ULVERSTON:
PUBLISHED BY D. ATKINSON, KING STREET.
LONDON:
WHITTAKER & Co., AVE MARIA LANE.
1869.

TO
A. BROGDEN, Esq., M.P.,
THE
LOCAL REPRESENTATIVE OF A FAMILY
THAT HAS
LARGELY CONTRIBUTED TO THE PROSPERITY OF FURNESS,
NOT ONLY IN DEVELOPING ITS MINERAL RESOURCES,
BUT IN
ESTABLISHING UNDER ALMOST OVERWHELMING DIFFICULTIES,
A COMMUNICATION WITH
THE RAILWAY SYSTEM OF THE KINGDOM,
THEREBY
EXTENDING THE COMMERCIAL RELATIONS OF THE DISTRICT
IN A REMARKABLE MANNER,
THE
FOLLOWING PAGES
ARE GRATEFULLY INSCRIBED
BY ONE WHO HAS RECEIVED A LARGE MEASURE OF HIS
KIND CONSIDERATION.

PREFACE.

THE following pages are presented to the public with many misgivings, and not until after long and serious thought. The author has yielded to the pressing solicitations of many kind friends, whose very flattering assurances have at last induced him to arrange in their present form, the results of a long lifetime of experience and observation, among the remarkable geological phenomena conspicuous in this peculiarly illustrative district. Any knowledge he may possess has been self-acquired, and amid his ordinary busy avocations he has contrived to snatch time for the prosecution of his favourite study, so that, with disadvantages unknown to his fellow-labourers in this part of the kingdom in point of learning, his work has been the more arduous and toilsome.

He has claimed the right of telling his story in his own humble way, otherwise he could not have been persuaded to assume the responsibility of authorship, and publish the accumulation of facts resulting from his long and personal acquaintance with every nook and corner of Furness. At the advanced age of 78 he cannot expect to see whether his contribution to the stores of his beloved science may be productive of any practical good to his native district, but he leaves his work with an indulgent public in the hope that his readers will give him credit for earnestness of purpose, and put down all his shortcomings in a literary sense to his inexperience as an author and to the overwhelming passion for a pursuit which has given intense pleasure during a privileged and lengthened existence.

Sedgwick Cottage, near Ulverston.

CONTENTS

PREFACE
INTRODUCTION ... 1
BOUNDARY OF FURNESS ... 9
STRATIFICATIONS OF FURNESS
   Green Slate and Porphyry ... 43
   Coniston Limestone ... 48
   Coniston Flag ... 60
   Coniston Grit ... 65
   Lower Ireleth Slate ... 70
   Ireleth Limestone ... 71
   Lower Ludlow Rock ... 78
   Mountain Limestone ... 87
   Upper Permian Sandstone ... 94

ITINERARY
   Duddon Bridge to Ulverston, Hawcote and Stank ... 97
   Pleasures of Geological Pursuits ... 101
   From Ulverston to Plumpton and the Seashore ... 104
   Ulverston to Birkrigg, Baycliff and the Seashore ... 108
   Ulverston to Swarthmoor, Urswick, Hawkfield and Gleaston Castle ... 114
   Ulverston to Lindal and Stainton ... 131
   Ulverston, Lindal Moor, Marton, Dalton to St. Helens ... 142
   Ulverston, Knotthollow, Iron Yeats, Groffacrag Scars, Gawthwaite to Kirkby ... 156
   Geological Wanderings in Cartmel ... 165
   Geological Rambles in the Lake District ... 178
   Geologising under Difficulties ... 184
   Geologising under Favourable Circumstances ... 221
   Further observations on the Haematite Ore of Furness ... 229
   Hoad and Outrake ... 233

ANALYSES OF IRON ORES ... 235

APPENDIX
   List of Fossils ... 244
   Elevation and Subsidence of Sea Coast of Furness ... 249
   Earthquake at Rampside, Barrow etc. ... 253

GLOSSARY OF TERMS ... 260
GAZETTEER ... 265
DRAMATIS PERSONAE ... 271

ILLUSTRATIONS.

	PAGE
Actinocrinus, from Knott-hollow Quarry, near Ulverston	*Frontispiece*
Section from Duddon Bridge by Ulverston to Stank...	97
Section of a Shaft at Lindal-Cote Mines, near Ulverston *(fig. 1.)*	132
Section of a Shaft at Lindal-Cote Mines, near Ulverston *(fig. 2.)*	134
Section of a Bore-hole at High Cross Gates	152
Actinocrinus Pulcher, from Knott-hollow Quarry ...	164

The fossil illustrations have been beautifully delineated by the artist from casts taken in gutta-percha from the originals by the author, who also drew the Sections.

INTRODUCTION.

The writer of the following pages has spent a considerable portion of a long life in "grubbing" amongst the rocks of Furness, Millom, and Cartmel, and from the early age of five years was placed in circumstances more than ordinarily favourable for geological pursuits. He now ventures to lay before the public the result of his experience in the quarry, on the mountain, and by the sea-shore, not presuming to set himself up as an instructor in the science of Geology, or to call himself a geologist, even of the humblest class, but merely a geological pioneer and guide to the various interesting geological phenomena which have come under his own observation.

Furness, from its insular situation, was almost inaccessible, and but little known, before the construction of the Ulverston and Lancaster Railway, consequently it was not much visited by geologists. Professors Phillips and King have made partial surveys of it, also Mr. Binney, of Manchester, an eminent geologist, but it is to the untiring zeal and perseverance of that venerable and humble christian minister, the Rev. Adam Sedgwick, Professor of Geology in the University of Cambridge, that we are indebted for the most reliable geological information of Furness, and quotations will occasionally be made, when requisite, from the works of the last-mentioned author.

INTRODUCTION.

Although Professor Sedgwick has conferred a benefit by his careful geological surveys of this district, yet it must be acknowledged that some little nooks and corners, or some important and unexpected organic remains, which may have escaped his notice, were more likely to be stumbled upon by a native observer in his more frequent geological rambles. Indeed, such has been the case, and one object of the author of this work is to endeavour to fill up a few of those little gaps left by the Professor, as omissions must unavoidably occur in all geological examinations made by strangers, especially in a complicated district like this, where the stratifications are broken up and contorted in every possible direction.

It may perhaps be interesting to the reader to know how or why the writer became a geological inquirer, especially as there was not a single geologist in Furness, at the early period he commenced his work; and, certainly, the first impulse given to his mind for the pursuit of geology was accidental. It is a remarkable fact that circumstances which are trifling in themselves, may and frequently do give to the human mind a bias for some particular pursuit, which, slight at first, might easily have been eradicated, but being fostered and unchecked, gradually gains strength until it becomes a passion, almost absorbing every other feeling; and although it may have been an occurrence of childhood, or school-boy life, yet it will cling to the individual, and exert an influence on his actions and pursuits in after life, even to extreme old age.

About the year 1795-96 the late Malachi Cranke, Esq., of Urswick (grandfather of the present Mr. Malachi Cranke, of Hawkfield), sunk a well in the south-east corner of a large rocky pasture field, called the "Hills," situated immediately behind and abutting on the ancient Free

School on Urswick Green. The material excavated from this well consisted of a mixture of clay, limestone, shale, &c., and contained innumerable fragments of organic remains, comprising portions of the columns, and single *osscicula* of several species of Encrinites, with beautiful sculptured plates of the head, a few fragments of fossil shells, and abundance of spines of a species of *producta*, coated with a bright silvery nacre. Some of this material is lying near the well to this day, but partially grown over. At that time the writer was living with his widowed mother and his younger brother, in a small cottage at the lower side of Urswick Green. So humble was this dwelling that the yearly rent was only 15s., and the cottage remains in the same state at the present time, with its flight of rough unjointed and unhewn limestone steps before the door.

One fine day, at noon, the writer wandered down to the well, and the sun shining brightly the while, caused a reflection of light from some object on the rubbish heap, which subsequent knowledge proved to be one of the silvery spines of one of the species of *producta;* this induced a further and more minute examination, when several fragments of different species of Encrinite came to view, consisting of portions of columns and also beautiful sculptured plates of the head. Daily visits to the place became the rule, and generally during play-hours an industrious search was made among the rubbish, always ending with a quantity of beautiful "fairy cheeses" in the pockets. The finders, like other philosophers, began to theorise on the subject of their formation, and in childish simplicity believed they were made by some clever workman, like "spinning tops" and other toys, but how or why put so deep in the earth, was a mystery which could not be unravelled.

One day, another little boy, a cousin, was prevailed upon to join in the search, and there might have been seen sitting, lying, or crawling about, a pair of "great philosophers" attempting to teach themselves on the rubbish heap at the well, the A B C of an almost unknown science, without any instructor whatever, while attending at school, the same were learning the A B C of the "scrapple book," under the able teaching of the late Rev. Wm. Ashburner, Vicar of Urswick. They had even begun to class and name the newly discovered beauties; all the silvery spines of the *productas* were called "pins and needles," and the single *osscicula* of the columns of the Encrinites "fairy cheeses," but for the beautiful sculptured plates of the head no specific name was adopted beyond the vague term of "queer things."

It is now seventy-one years since the events here recorded took place, and that childish companion was until very recently alive, and residing at Gleaston. He was, like the writer, an old man, but he well remembered geologizing at the well, and talked with pleasure about the "pins and needles," "fairy cheeses," and "queer things," which puzzled his understanding. After a conversation on the subject one day he said, rather sorrowfully, "after you left Urswick and went to live at Ulverston, our companionship seemed broken up, and I could not bear to go to the old place by myself, for although I gave up geologizing from that time, whenever I go near the scene of our boyish days it brings to my mind a flood of recollections of our early amusements which are of a pleasing but melancholy nature, and I have sometimes blamed myself for not persevering as you have done through the whole of your life. I have no doubt it gives you pleasure yet, otherwise you would not continue your geological rambles on the mountains, and expose yourself to dangers of no ordinary kind." "You say truly, cousin James," was the reply, "it does give me pleasure, as it always has done, such as no earthly pursuit ever did or ever can do, and I humbly ask the Lord that so long as I am spared with eyesight to read His word in the sacred Book, so long may I possess my other faculties and be enabled to read the wonderful record of His works engraved in the rocks and stones of my own beloved Furness."

It will be seen from the above simple narrative, that the author's work as a geological pioneer, commenced several years before that highly gifted geologist, the late Hugh Miller, of Edinburgh, was born; but he does not presume to compare himself in the remotest degree, with that wonderful man, only so far as the fact that both had to grope in the dark without instructor of any kind, — one as a mason's apprentice in a small country town in Scotland, and the other as a very young school-boy at Urswick, the utmost extent of whose learning did not reach beyond the alphabet,—indeed this sketch or fragment would not have been written but for two reasons; one of which, although personal, may be permitted from its intimate connection with the other.

The first is to show that the trifling incident of a reflection of the sun's light from a heap of clay and stones, should be the starting point of a career of rambling amongst rocks and stones, which has continued with increasing interest and pleasure for upwards of seventy years, and will last until the Lord in His own good time shall call the writer hence.

The second reason for writing involved one of the strangest geological facts that the mind can conceive, showing that the infant science of geology has not progressed in accordance with the other natural sciences, and that geologists have yet very much to learn. It will scarcely be believed that some of the "pins and needles," "fairy cheeses," and "queer things" of the author's childhood are declared to be new even at this day, and not only are there several new species, but amongst the "queer things" (sculptured plates of the heads of Encrinites) there are two or three also which are of new *genus*, and entirely unknown to science. From this it may be inferred that the very humble cottage at Urswick, was a geological museum in miniature, and contained many unknown and beautiful organic remains, of which neither the British Museum nor the Government Museum of Practical Geology, in Jermyn Street, can boast, although they were displayed as children's play-things in that cottage so many years ago.

In proof of this it may be stated that the late John Brogden, Esq., F.G.S., of Lightburne House, Ulverston, a short time before his lamented death, took up to the Museum of Practical Geology, and also to the British Museum, a few cards of beautiful organic remains of the same species as those collected at the early period already mentioned. These were shown at the British Museum, and also to Mr. Salter, Palæntologist to the latter institution, a gentleman of the highest scientific attainments in that particular study, who declared that several of them were of new species, and two or three of new *genus*, and absolutely unknown to science. He marked them at the time and a few of them are still in the writer's collection, but the greater portion are in the cabinet of Alex. Brogden, Esq., M.P., of Lightburne House, Ulverston. Amongst these are several fragments claimed to be portions of the *tentacula* of the Lily Encrinite of Bohemia, which have never before been found in Britain. These fragments, however, are, on the high authority of Mr. Salter, and Mr. Woodward, one of the curators of the geological department of the British Museum, denied to be the Lily Encrinite; nevertheless, as these gentlemen were not at the time able to name them, and as the minutest examination, on the author's part, cannot discover the least difference between these fragments and the true Lily Encrinite, until some evidence be found to the contrary, they may *provisionally* be called by that name.

The reluctant removal of the writer to Ulverston at the age of nine, resulted in his having to labour hard from morning to night in a weaving-shop for eighteen-pence a-week, and although he had but little spare time allowed, he managed in a few years to visit every mountain limestone quarry in Furness, and some of them several times, not forgetting to devote a holiday occasionally, (whenever good luck brought one,) to the favourite spot behind the Free School on Urswick Green.

Frequent visits to Plumpton quarries, revealed the different class of fossils there prevailing, viz., *cornulites, spirifer, orthis, terebratula, &c., &c.*, and along the shore from the quarries northwards to Plumpton Hall, several species of corals imbedded in the smooth sea-washed floor of limestone rock, the original colour of many of them being still perfect, but quite impossible to obtain without "blasting." There are also numerous fragments of Encrinital remains imbedded in the cliff, consisting principally of portions of columns or stems, but their markings are obscure and they are utterly worthless.

During boyhood and early manhood the sea-shore was a favourite resort, particularly from Conishead Bank, by Wadhead Scar, Bardsea, Bienwell Scar, to Aldingham;

and many a heavy load of organic remains was carried home from the scars along this shore; all of which were prized as treasures, though at that time unknown. Amongst those found on the shingle opposite Bardsea Mill, a good specimen of *catanipera* or chain coral occurred, and several specimens of coral showed, when polished, interior organization more perfect than any fossil the author has ever yet seen. A polished slice of one of these fossils is now in the British Museum, the remainder are in the cabinet of Alexander Brogden, Esq., M.P.

Birkrigg and Baycliff quarries were often visited, and St. Helens, near Dalton, was prized for one particular species, the *nematophyllum arachnideum, n. minus*, of which were obtained several splendid specimens.

Besides the places of interest here enumerated there are several others to be more particularly described hereafter, but as this was not intended to be an autobiography, but an introduction to a Physical and Geological History of Furness, it will be necessary to describe the boundary of the district under consideration.

Ulverston, January, 1869.

GEOLOGICAL FRAGMENTS.

BOUNDARY OF FURNESS.

The Hundred of Lonsdale North of the Sands, in the county of Lancaster, is entirely detached from the rest of the county by Morecambe Bay and a portion of the county of Westmorland; so that geographically it may be considered a part of Westmorland.

Lonsdale North is divided into Cartmel and Furness, there being a good natural boundary between them— Ulverston Sands, the River Leven, and the Lake of Windermere. It is of Furness, or the western portion of the Hundred, we are about to speak. Furness is a long, narrow district, bearing nearly due north and south; its extreme length from the Isle of Walney, on the south, to the nab on the south shore of Elterwater, on the north, is about 25 miles, and averaging from six to seven miles in breadth from east to west. The southern portion of it, for ten or twelve miles, is a promontory running down between Morecambe Bay and the estuary of the Duddon, and terminated by the Island of Walney, at the extreme point of the promontory.

The district is bounded on the east by Morecambe Bay, the River Leven, and the Lake of Windermere; on the north-east, by Westmorland; on the north-west and west, by the county of Cumberland; and on the south by the Irish sea.

We will now attempt a perambulation of the boundary in imagination, as indicated above, having made the same in reality more than once on foot (which is the only way in which it can be accomplished). We shall also slightly notice some of the geological phenomena of interest in our way as we pass them, commencing at the Lighthouse at the south end of the Isle of Walney, passing Piel Castle, and Rampside, by keeping to the shore of Morecambe Bay, we come to the great shell bed of Roose Beck. The shingle here consists principally of granite and mountain limestone, and on the shore opposite Newbiggin shingle and large pieces of very hard new red sandstone occur, containing numerous vegetable fossils. This sandstone is of a different type to that of Hawcoat and Furness Abbey, which are the only places in Furness where new red sandstone or permean rock, of Murchison, can be seen in *situ*, and in neither of these places have fossils been found. Hitherto we have seen no rock of any kind in *situ*, either on the shore or inland. Proceeding northwards limestone greatly predominates in the shingle on the beach, and near Aldingham — five miles from our starting point — we come to the great and beautiful mountain limestone deposit of Furness, uninterrupted on the shore northwards for a distance of six miles, and continuous westwards across the whole of Low Furness to the estuary of the Duddon. After leaving Aldingham, and crossing Kirk Scar, which abounds with corals and other characteristic mountain limestone fossils, we direct our steps northwards for about a mile to Bien Well (good well), a copious spring of very pure water, on the beach, yielding about 500 gallons per minute. We are now opposite the village of Baycliff, and distant from it about half a mile there are two quarries in work, which produce beautiful and almost pure white stone, susceptible of the very highest polish; these, and the quarries of Mr. J. Garden, at Stainton, two miles west from Baycliff, yield, we believe, a stone superior to that of any mountain limestone in the kingdom, particularly for church and other ornamental work where delicate tracery is required. [The author has paid considerable attention to this subject, and, although, he has visited most of the limestone quarries of England, he has never seen any stone equal to it.] We now return to the sea-shore, at Bien Well, where there is a splendid floor of limestone dipping to the east at an angle of about fifteen degrees and extending northwards for a quarter of a mile; it is smooth, white, and clean, being washed by every tide, and shows a great many species of organic remains, consisting of the *genus producta*, *spirifer*, *euomphalus*, *astrea*, corals, &c., most of them being in relief owing to the unequal weathering of the rock, and when left wet by the receding tide some of them show interior organization. A mile further north we come to Sea Wood, at which place the late T. R. G. Braddyll, Esq., found copper ore, but not in sufficient quantity to be worked profitably, and the mines were given up. At Sea Wood there is a quarry from which a considerable quantity of limestone has been shipped to Liverpool and other places, wherein very good specimens of the *edmondea sulcata* have been found, a fossil of the type of the common mussel, which is rather scarce in the mountain limestone. Here, also, is a beautiful floor of limestone, which reaches to the north end of Sea Wood, a distance of a quarter of a mile. From this point we have no rock on the shore for about a mile; but a little way inland, at Well Wood House, the residence of Mrs. Petty, there is a private quarry, which abounds with the *producta gigantea* and other fossils. Immediately behind the grounds rises that immense mass of mountain limestone, the hill of Birkrigg, nearly 500 feet in height, and covering an area of 350 acres. We now pass the ancient village of Bardsea, and the quaint and picturesque Bardsea Hall, the seat of Capt. R. H. Gale, J.P., where a beautiful natural rockery may be seen in the gardens, and in the park behind an interesting terrace of limestone rock surmounted by a cenotaph, in memory of the Gale family. Our next object of geological interest is Wadhead Hill, a bluff headland jutting out some distance into the bay, to which it presents an almost perpendicular escarpment fifty feet in height. This hill is almost entirely composed of sand, gravel, and other water-washed material, and, notwithstanding it has been protected by a very strong and well cemented wall, is rapidly wasting by the sea, every high tide with a strong south-west wind brings down and carries away large portions of it, strewing the scar in front with innumerable fragments of petrified sand of the most fantastic forms. Near the base of the hill there is a perfect horizontal stratification of these petrifactions, very hard, durable, and exceedingly ornamental, and when detached from the hill by storms, the best portions are collected and used for ornamental rockery. The hill slopes rapidly inland, and in time will be washed away entirely. Half a mile north of Wadhead Hill is Conishead Bank, which was, within our recollection, the principal place for shipping the hæmatite iron ore of Furness. We are now opposite Conishead Priory, in the grounds of which the mountain limestone is developed on a great scale, but from Conishead Great Head Wood it dips rapidly to the north-east, and is soon covered up in the low lands to the north, forming a junction with the clay-slate in that direction about a mile from this point.

Hitherto we have had a fine clean shingle beach during the whole of this long traverse — nine miles — but it ends here rather abruptly, and, except two or three small patches, there is no more shingle on the western shore of Morecambe Bay. Pushing forwards about a mile, we come to the limestone deposit of Plumpton, where there are two quarries in work, one of them is on the shore, and produces a fine white stone, the other, is about one hundred yards further inland, and yields a stone very dark, but highly fossiliferous. There is here on the beach a fine floor of smooth white limestone similar to that at Bien Well; washed by the sea and displaying innumerable organic remains, some of them showing their original colours, and interior organization. This floor is continuous to Plumpton Hall, where it forms a junction with the clay-slate, not exhibiting the actual contact of the two rocks, but in the field behind the farm-buildings, both these formations are seen on the surface within a short distance of each other. This clay-slate is the beginning of that immense deposit of slate rock which Professor Sedgwick divides into three members, *i. e.*, the Lower and Upper Ireleth slate, and the Lower Ludlow rock; the Upper Ludlow, and the old red sandstone, which should follow in sequence, are wanting, therefore the mountain limestone comes down upon the Lower Ludlow rock.

From Plumpton Hall northwards no more mountain limestone occurs in Furness, except a small patch at The Ashes, being an outlier, only covering six or eight acres; entirely surrounded by clay-slate. On the opposite side of Ulverston sands there is a fringe of limestone from Holker Park, by Capeshead, Parkhead, and Low Frith, to Hazlehurst Point, where it forms a junction with the Lower Ludlow rock; it then leaves the sea-shore, inclining inland, and at the distance of two miles, ends with a small point at the north end of Roudsea Wood. It is highly probable that this fringe of limestone is continuous under the sands from Holker Park to the south-west by the Black Scars, Chapel Island, and Elwood Scar, to the north end of Seawood; therefore it forms a part of the great limestone deposit of Furness.

On the west side of Holker Park, the seat of His Grace the Duke of Devonshire, there is a most instructive natural section of rock, showing the junction of the mountain limestone and the lower new red sandstone, some parts of it abounding with fossils characteristic of the coal formation. From Plumpton we have a soft clay beach to Tredlea Point (about 500 yards) where the clay-slate crops out abruptly and forms the western abutment of the Leven viaduct on the Ulverston and Lancaster Railway. For this railway which has done so much and so well for the development of the immense mineral resources of Furness, and given a splendid impetus to the fortunes and prosperity of the Borough of Barrow, in Furness, we are entirely indebted to John Brogden, Esq. and his family. There is no example of a similar kind in the United Kingdom, where a single private family has undertaken, and with energy and perseverance completed, a difficult and hazardous engineering work, involving an outlay of four or five hundred thousand pounds, to be derived principally from their own resources, and this in the face of opposition and prejudice, may well claim our admiration and respect.

It is somewhat startling to a stranger when crossing the sands by railway for the first time, at high water, to observe vessels sailing on both sides of him, and a few minutes after passing over the viaduct to see a portion of it open, and the ships pass through.

From Tredlea Point, half a mile north we come to Ashes Wood, the small patch of limestone above alluded to. Here is a small private quarry, producing very fine grained stone, used as a flux for smelting iron ore at the furnaces of Messrs. Harrison, Ainslie, & Co., at Newland and Backbarrow, (this is the only firm in England which engaged in smelting iron ore with wood charcoal, and the iron produced by this process is said to be the best in Britain.) From the provincial name of the stone (bloomery) it was very probably used for the same purpose at the ancient bloomeries of High Furness, the remains of which may be found in the neighbouring hills. This quarry is the last place northwards where mountain limestone is seen in Furness, and very few organic remains have been found in it.

We now enter on that immense deposit of clay-slate which extends towards the north for twelve or thirteen miles without much perceptible geological change.

Two miles further on is Greenodd, a small seaport, where there is a quarry in the clay-slate, the stone not being good for building purposes, is principally used for road metal, and no organic remains have been discovered in it.

Ulverston Sands, or the western branch of Morecambe Bay, ends here, but the tide flows to Low Wood, two miles further up the river Leven. Soon after leaving Greenodd, we pass Penny Bridge Hall, the seat of J. P. Machell, Esq., J.P., through whose gardens and grounds runs the river Crake, the outlet from Coniston Lake, and forms a junction with the Leven, the outlet from Windermere. This "meeting of the waters" is quite equal in beauty to that in the "Vale of Avoca," celebrated by Thomas Moore, in his song of that name. Both these rivers run their whole course in the Lower Ludlow rock, which here juts out with a bold escarpment almost into the head of the Bay. Some portion of the extreme point was removed not many years since, for the purpose of making a road between the rock and the river Leven.

Our course is now by the Leven, through extensive

mosses, in which are embedded the trunks of large trees, and on the opposite side of the river, at the north end of Roudsea Wood, there are the ruins of a forest covering some acres of land; the trees are not uprooted but broken off some height above the surface, their roots firmly embedded in the ground. This moss-land is a good botanical field — several rather scarce plants have been found upon it. The marsh fern *(lastrea thelepteris)* grows in two or three places, most luxuriant specimens of the royal fern *(osmunda regalis)* may be seen, and the sweet-scented gale *(myrica)* is very abundant. Several horns of a very large species of deer have been dug up from a considerable depth in the moss; most of which are now in the possession of His Grace the Duke of Devonshire, who is owner of a great part of the surrounding district.

About a mile further is Birk Dault, the seat of the Misses Barker, a short distance from the Low Wood Powder Mills. Hence to the foot of Windermere Lake the Leven runs almost entirely in a rocky channel, particularly at Backbarrow, where huge masses of stone crossing the bed of the river make a considerable fall, almost deafening in floods, and notwithstanding this great obstruction, salmon are able by some means to surmount it in the breeding season, and pass up the river into Windermere. At Backbarrow are the cotton mills of T. Ainsworth, Esq., not like factory buildings in general — eyesores and nuisances — but on the contrary rather an ornament to the scene. One mile up the river is Newby Bridge, at the foot of Windermere, with its first-class hotel for lake visitors, from which steamers ply during the summer season.

From Newby Bridge to the Ferry, six miles, there is no very marked geological change, neither are there any geological phenomena calling for particular remark. We are on the Lower Ludlow rock and the Upper and Lower Ireleth slate the whole way. In this long traverse, keeping near the west shore of Windermere Lake, we pass the Landing, High Stott Park, Graythwaite, Cunsey, crossing Cunsey Beck (the outlet from Esthwaite Lake), which falls into Windermere at this point; and about a mile beyond, near the Ferry Inn, we reach a quarry where several species of organic remains have been found. From the Ferry northwards, the Lower Ludlow seems imperceptibly to change into the Coniston flag, and must have already blended with the Upper and Lower Ireleth slate, there being no perceptible difference between them in this part of their range. On the Crier of Claife, two miles north of the Ferry, is a quarry in which fossils have been found, and Cold Well quarries, two miles north from Claife, have produced many splendid specimens of the *obtusa caudata*, and other characteristic fossils of the Coniston flag. Half a mile north of Cold Well are the celebrated quarries of Brathay, which we believe to be the best of their kind in England, for the Coniston flag formation is seen in all its beauty. Large and excellent flags are produced here, and a considerable quantity of the stone is sent to distant parts of the kingdom for ornamental purposes, such as edgings for garden walks, &c. They are about two inches thick, self-faced, very true and straight, generally ten or twelve inches broad, with square edges, coated with iron pyrites, beautiful as gold; many of them ten or twelve feet in length, and resembling deal planks. These quarries have produced a few species of organic remains, especially the *graptolites ludensis (priodon)*, usually found in the nodular concretions with which the workings abound. A little way north of Brathay quarries, at Pull Wyke, we come to that eminently fossiliferous formation, the Coniston limestone (the equivalent of the Bala limestone in Wales), a long narrow band, commencing near Beck, in Millom, with a north-east strike for upwards of fifty miles. It enters

Windermere at this point and appears again on the east side of the lake near Low Wood, in Westmorland. This thin band of limestone does not average more than 200 or 300 feet in thickness, and at one particular place (Broughton Mills) not more than 30 feet. It abounds with many rare species of organic remains, several undescribed fossils, some of which are unknown to science. This is the lowest stratification in Furness where organic remains have been obtained.

At Eagle Crag, near Brathay Garths, we come upon the green slate and porphyry, running thirteen or fourteen miles in a northerly course without any change whatever. After passing Brathay Hall we arrive at Brathay Brig, the extreme north-east limit of Furness, a little below which we have the confluence of the rivers Brathay and Rothay, the united stream entering Windermere about 600 yards lower, at Gale Nase Crag. There is here a curious problem for our friends the naturalists to solve, — in the breeding season the salmon and the char proceed in company from Windermere to this "meeting of the waters," where they bid each other good-bye, the salmon says, "I will make my bed in the Rothay," the char says, "I will make mine in the Brathay;" they part, with friendly greetings, to meet no more until early spring, when they descend the united stream in company to Windermere, the char to remain in the lake, the salmon to pass through it and down the river Leven into the sea. These two rivers are equally pure, both having gravelly, and in some places rocky bottoms, indeed they are to all outward appearance the same; for, like two beautiful twin sisters, we scarcely know one from the other, and whatever condition is favourable for breeding in one river is common to both, yet the salmon and char select each its own river, not mixing with each other or changing breeding ground from generation to generation.

BOUNDARY OF FURNESS.

From Brathay our course is due west for about a mile and a half, by Jeffy Knotts Wood and Hunter How Coppice to Skelwith Bridge, thence north-west one mile further by Skelwith Force and Rumple Crag to the Nab on Elterwater. This is the extreme northern point of Furness, therefore the old saying "the greatest length of Furness is from Bratha Brig to the Peel of Fudder," is not quite correct, its utmost extent being from the south end of Walney to the Nab of Elterwater. This Nab is also the extreme northern part of the county of Lancaster, which runs up many miles between the counties of Cumberland and Westmorland, and ends here with a very small point, like a wedge driven in between them.

We may also state for the interest of our geographical friends (who are very particular in matters of description), that this point is situated in latitude 54° 25′ 44″ north, and 3° 1′ 30″ west longitude.

Returning about a mile, nearly south to Colwith Bridge, thence due west again to Colwith Force, by High Park, and Stang End, to Low Hall Garth, about a mile and a half, we have Black Hall, Moss Rigg, and Wood slate quarries. Half a mile south we have Betsy Crag, Tunnel, Sannibuts, Mirk Hall, and Sty-Rigg quarries, besides which there are others southwards, all the way to Tilberthwaite. These quarries are all in the green slate and porphyry formation.

Our course is still due west, passing Little Langdale Tarn, Bridge-end, to Vickers—about a mile. At Vickers we begin to ascend Wry-Nose mountain, to the Shirestones, or three-foot Brandreth, a mile and a half, where the counties of Cumberland, Westmorland, and Lancashire meet. Here are also the spring-heads of the rivers Duddon and Brathay, giving a good natural boundary to Furness, which is entirely surrounded by water. The river Brathay is the boundary eastward to Windermere, thence south by the river Leven, to the sea; the river Duddon is the boundary south-west to Seathwaite,. then south to the estuary of the Duddon and the sea. We are here on a table-land 1270 feet above ordnance high water mark, and about 1000 feet above the valley of Great Langdale. Although this may be considered the summit of drainage, yet it is, in fact, the south end of a mountain valley, the north end of which falls into Oxendale, a branch of Great Langdale. On the east side of this valley rises the Pike of Blease, and from it issue three or four porphyry dykes; on the west, Cold Pike and Cringle Crags, and between them lies Red Tarn, at the north end of which — one mile from the shire stones — there is a great deposit of iron ore on the surface, covering ten or twelve acres, or perhaps much more. Some of the ore is not good; but the greater part of fair average quality. This was the site of Mr. Cram's mining operations, which are now given up on account of the difficulty and expense of getting the ore down into Oxendale, and thence into Great Langdale, as it would require two steep inclines, one 600, the other 1200 yards in length, with a gradient of one in four and three-quarters. We think this ore could be wrought profitably by a company who would spend eight or ten thousand pounds in preliminary operations. The ore lies on the surface, and, as the land has sufficient fall into Oxendale, neither wooding nor pumping would be required; these are considerable items of expense in all mines, where they are necessary, and, we think, would balance the expense of forming the inclines, and other constructive works. From Cringle Crags descend three becks, each down its own ghyll, all difficult to traverse, particularly Hell-Ghyll, which is almost entirely inaccessible, and really dangerous.

As we have already stepped two or three miles out of Furness, we will attempt to describe this strange place before we return. Hell-Ghyll, when viewed from Red Tarn, appears an immense deep and gloomy crack in the mountain on the opposite side of Oxendale, it is wedge-shaped, the point coming down almost to the foot of the mountain, widening and deepening as we ascend for about half a mile, when its progress is abruptly ended by a perpendicular face of rock of great height, which entirely crosses this gloomy chasm. In a conversation with our shepherd friend, living at Wall-End, in Langdale — the nearest house to the ghyll — who is an active and daring young man, generally going down the rocks with the crag-rope around him, when sheep are crag-fast, (whether the sheep are in his own care or belong to his neighbours); said he had never been in Hell-Ghyll, and was not acquainted with any one who had. This circumstance excited our curiosity, and we determined to make the attempt. Having prevailed on a young miner to accompany us, with long crag-staves in our hands we entered the ghyll from the lower end. Proceeding a short distance we soon discovered the difficulties to be surmounted, and we had to use our staves constantly to help us up the rocks, (which entirely filled the bottom of the ghyll, and were most of them wet and slippery) and although the brook which runs through it was difficult to ascend, we found it still more arduous to return by the same way, we therefore went boldly forward, hoping to find our way out at the other end. Our progress however, was very slow, and we found we had begun our journey too late in the day, only three hours before sunset, and that we should not have sufficient time to examine it so carefully as we otherwise would have done; besides, the nature of the place makes it gloomy at all times, and as the evening drew on, it became even more so, notwithstanding which we discovered two or three veins of iron ore, as well as a piece of lead ore two inches in daimeter, lying in the bottom of the ghyll. We also observed some rather scarce plants and ferns, particularly the *Hymenophyllum Wilsonii*, which grows on the wet rocks in two or three different places. The iron ore here, as well as at two other localities in a neighbouring ghyll, which comes down from Cringle Crags, two spots at Langdale Pikes, and at all other places in the green slate and porphyry formation that we have seen, is in vertical veins, unlike the hæmatite ore of Furness, which lies in sops or great amorphous masses. We now began to examine the cliffs on both sides of us, to discover where we could climb to the surface, but found that was impracticable, and at length our progress was completely stopped, the ghyll which had been gradually widening from the commencement, was here terminated by a wall of rock running across its widest part (as already described), in fact, it ended abruptly but entirely, the surface being no longer a ghyll, but a comparatively smooth mountain side, with a gentle depression for the beck course. We had to retrace our steps, with darkness coming on, and as it was impossible to find a dry nook amongst the rocks for a lodging during the night, we again examined the cliffs on both sides more minutely, and at length selected a spot where the rock was not quite so steep, but loose in some places; agreeing between ourselves that one of us should make a trial to ascend, while the other remained at the bottom, keeping out of the way of loose rock, which was certain to fall. However, although we found the ascent both difficult and dangerous, we both safely arrived at the top of the cliff, truly thankful, and would not like to attempt the same feat again. We would not have mentioned this ghyll in any way, only that it involved a geological difficulty of some interest. As we have before stated, it is in the form of a wedge, with the broad end up the mountain, the length twelve or fourteen times its greatest breadth, with solid rock on both sides and end, the indentations and

BOUNDARY OF FURNESS.   23

projections of the sides appearing to correspond. As Scawfell (of which Beaufell once formed a part) is supposed by Professor Sedgwick and other eminent geologists to be the centre of disturbance which has fractured and broken up the whole of the lake district, — Wastdale, Eskdale, Seathwaite, Oxendale, Langdale, and Borrowdale, are some of its effects, all of them radiating from it, like the spokes of a wheel from the centre—and looking at the great extent of effects, it is evident the cause of disturbance was deeply seated in the earth's crust, and that these are the primary cracks resulting from such disturbance. It is probable, however, that the secondary cracks, or ghylls, were not produced in this way, and some are undoubtedly due to the mountain streams that rush through them. There are two of these coming down from Cringle Crags which have a very small beginning, and gradually deepening as they descend the mountain, are augmented by others falling into them. This is not the case, though, with Hell-Ghyll, which is widest and deepest at the upper end, and gradually becomes narrower and shallower as it descends, until it reaches a point nearly at the foot of the mountain, so that when viewed from the opposite side of Oxendale, about a mile distant, where the eye can take in the whole at one view, presents the appearance of having had a large acute triangular piece cut entirely out and removed, not as if some great convulsion of nature had opened a crack in the side of the mountain, which another convulsion under different circumstances could close up, but from its shape we may conclude that no combination of forces, however powerful, could close it up again. Neither does it furnish any conditions or data for correctly interpreting its formation. We can only say it is a mystery and leave it.

We now return to the Shirestones on Wry-Nose, the source of the Duddon; which we follow south-west for two miles to Cockley Bank, where a copper mine was wrought a few years since. It was not a profitable speculation, but the ground has not been thoroughly proved, although there are good indications of copper. A little further we enter the Seathwaite valley (Wordsworth's own sweet vale of Duddon), where the Duddon receives a principal tributary from Hard-knot, and turns abruptly to the south, by Black Hall, one of the largest sheep-farms in the district, thence by Dale-head and Hinning-house to Wegbarrow Point, Gold-rill, Gold-rill Dub, Trout-Hall, Throng, Nettleslack, Holling-house, Undercrag, Turner Hall, to Seathwaite Parsonage (once the home of the wonderful Rev. Robert Walker), to Newfield, a distance of four miles. Here is a splendid porphyry dyke, rising with a sharp edge from the bed of the Duddon to a great height. South-west by Hall Dunnerdale is to Kiln Bank, and near to it on the Cumberland side of the river, Commonwood quarries (Messrs. Postlethwaite). The slate and flags produced here contain a most instructive geological lesson, illustrating faults, breaks, waves, and contortions of every possible kind, the lines of deposit being as clearly defined as if drawn with a pencil. Some of the stones are now used for mantle pieces, and other ornamental purposes, and are very beautiful.

From Kiln Bank, to Ulpha Kirkhouse, and Whistleton Green, three miles, the Duddon then turns due south by the Ancient Bloomeries at Cinderstone Beck, near to which two porphyry dykes (discovered by Professor Sedgwick several years since) are seen in the bed of the river. Between Cinderstone Beck and Stonster, high up the hill, are some ancient stone walls, and nearly on the top of Dunnerdale Fells, are a great number of ancient cairns. From Stonster by Whineray Ground to Rawfold, one mile, opposite Rawfold on the Cumberland side of the river, is Duddon Hall, the beautiful seat of Major W. S.

Rawlinson, J.P. From Rawfold by Bankend to Duddon Bridge, one mile. Our long traverse from Gale Nase Crag, where the river Brathay falls into Windermere, westward to the Shirestones, south-west to Seathwaite, south to Duddon Bridge, a distance of upwards of twenty miles, has been entirely on the green slate and porphyry. We are now about to cross several stratifications in our journey southwards, like crossing the "rigs" in a ploughed field. The first of these is the Coniston limestone, which first crops out near Beck Farm, in Millom, thence on the west side of Millom Castle, by Waterblain, Graystone House, to Duddon Bridge. From this point it takes a general north-east direction, to near the north end of Windermere Lake, and if a straight line be drawn from Duddon Bridge to Brathay Bridge, it would nearly represent the boundary between the Coniston limestone and the green slate and porphyry, all that great and wild tract of country on the north-west side of the line belonging to the latter formation.

For those who have not made geology a study, it will be necessary to enumerate the different stratifications of Furness, and the order in which they were deposited. According to Professor Sedgwick, they are as follow:— Green slate and Porphyry, Coniston limestone, Coniston flag, Coniston grit, Lower Ireleth slate, Upper Ireleth slate, Ireleth limestone, Lower Ludlow rock, Upper Ludlow rock, Old Red sandstone, Mountain limestone, New Red sandstone, and Magnesian limestone. Of these we have a perfect sequence except the Upper Ludlow rock, which is wanting in Furness, the nearest being Benson Knott, near Kendal.

In traversing the eastern boundary of Furness, from the south end of the Isle of Walney, to the mouth of the river Brathay, as already described, it is possible we crossed the whole of them in descending order, *i. e.*, from the newest to the oldest, but the boundary between each is very uncertain, because there is no rock on the surface for the first five miles, and all the slate formations seem to blend imperceptibly into each other. But in our course from Duddon Bridge to the north end of Walney, the different stratifications are more clearly defined, and as the Coniston limestone ranges across the whole of Furness, it is considered a very good base for a natural section, and its line of strike, from Duddon Bridge to Brathay, being north-east, and the other formations parallel with it, they have a general north-east direction. It is evident, therefore, in travelling from Duddon Bridge to the north end of Walney, we shall come athwart them like ploughed ridges, but very unequal in breadth ; and as we travel the eastern boundary from the highest stratification to the lowest, so in journeying over the western boundary we rise in the series from the lowest to the highest, and from the oldest to ths newest. With these preliminary observations we will proceed.

After leaving Duddon Bridge we enter on the Coniston limestone immediately, without seeing it, there being no rock on the surface, but it ranges up the valley of the Lickle, on the east side of Bleanslea Bank, and from its thickness near Hartley Ground on one side, and Graystone House on the other, it probably does not exceed 300 feet where we cross it on the road to Broughton. The Coniston limestone is a dark blue, hard, compact rock, possessing considerable hydraulic properties, and will harden under water. It is not much used for building purposes, being subject to rapid decay, examples of which may be seen in the crumbling walls on Applethwaite Common, between Troutbeck and Kentmere. These walls would yield a rich harvest to fossil collectors, there being many rare species of organic remains. But to return, before we arrive at High Cross, we enter on the Coniston flag, which continues through Broughton to Wall-end, where it forms a junction with the Coniston grit. The Coniston flag ranges across the whole of Furness, parallel with the Coniston limestone, and is a highly valuable stone for every purpose for which it can be applied,— a beautiful section of it may be seen behind the Railway Station at Coniston. Besides this formation producing slates and flags, it is extensively used for building, and a fashion now prevails, in erecting good houses, to use true self-edged stones from two to four inches in thickness, laid with superior mortar — the mortar not applied flush with the front of the wall, but laid back five or six inches from it so that no lime of any sort is seen on the outside of the building,— this has a very good effect. The Coniston flag is sparingly fossiliferous. The Coniston grit, as seen at Wall-end, Borderrigs, Wreaksend, Foxfield, and in the greater part of its range, is a hard, intractable, fine grained grit stone, of a reddish-brown colour, and has not been much used for economical purposes, their being no necessity for it, the superior Coniston flag being so near. This grit is almost entirely without organic remains, and ranges with the Coniston limestone and the Coniston flag to the north-east, but ends entirely at Foxfield to the south-west, whereas the other two formations can be traced for several miles in that direction across the Duddon into Cumberland.

In one geological ramble we discovered a porphyry dyke on Wall-end Moss, highly micaceous, decomposing into small shining flakes of a golden colour, traceable a considerable distance to the north, and ending abruptly under the moss. We next cross the great moss of Angerton, about a mile in breadth, and come to Causeway End, and are now on the Lower Ireleth slate; the junction between this and the Coniston grit being covered by the moss. This stratification skirts the low land by Bankend, Foul Bridge, Dove Ford, Dove Bank, Chappels, Kirkby Hall,

Wall-end, Headcrag, and Sandside, where a large quarry has once been wrought, but now given up; indeed it is not likely any more will be opened in this formation, as we have an inexhaustible store of superior slate in those immense quarries in the Upper Ireleth slate, about a mile east from Sandside. These splendid quarries are known to geologists in all parts of the world, principally from the untiring labours and writings of Professor Sedgwick, of Cambridge.

Professor Sedgwick subdivides our slate formation into four members — *i. e.*, the Lower Ireleth Slate, the Ireleth Limestone, the Upper Ireleth Slate, and Lower Ludlow Rock. We believe there is no well-defined boundary between the Lower and Upper Ireleth slate, but that they blend imperceptibly into each other, and as we ascend from the lower formation at Sandside towards the great quarries high up the ridge to the east, the slate metal has more perfect cleavage, is altogether a better material, yet has the same mechanical texture, and the same chemical composition, and upon the crown of the hill on Kirkby Moor, where the Lower Ludlow rock may be supposed to begin, the cleavage gradually becomes more and more imperfect to the eastern boundary of Furness. Proceeding southward, we pass Soutergate, Bailiff Ground, and Brighouse, to High Mere Beck, where we come on the Ireleth limestone of Professor Sedgwick, but as it is almost without organic remains, it cannot without further investigation be determined an outlier of Coniston, or an outlier of mountain limestone.

One mile west from Mere Beck, is Dunnerholme, a remarkable bluff headland of mountain limestone, which juts out into the Duddon estuary, and is almost entirely insulated by the tide. The quarries here abound with characteristic mountain limestone fossils. On the west side of Dunnerholme, where the rock is washed by the sea, there is a beautiful bed of red limestone, about eighteen inches in thickness; the beds above and below it are a good white. Continuing southwards by Tippin's Bridge and Moor Side, we have the Ireleth limestone exposed in a few trial quarries, but it ends entirely near the latter place. It has no great thickness in any part of its range, and as we trace it northwards, it gradually thins out, and ends with a small point in the upper part of the Low Hall estate, near Beck Side. It appears again on Tottlebank Fell, in two or three small patches, but they are worthless. A small patch was also pointed out to Professor Sedgwick and Dr. Gough, on Kirkby Moor, nearly in the same range, and somewhat similar in character, by Mr. Edward Coward, of Gill House, a very shrewd observer, who is in possession of some valuable facts for the antiquarian.

In our traverse from Bank End to High Mere Beck, a distance of three and a half miles, through a low and comparatively level part of the parish of Kirkby Ireleth, we have been entirely on the Lower Ireleth slate; but high up the ridge of Kirkby Moor, a mile to the east, the Upper Ireleth slate ranges nearly parallel with us, and seems to end with the ridge of hills at "Rebecca Quarry," near Moor Side. The whole of this high ridge is sparingly fossiliferous, several trial quarries having been opened in the southern part of it, but all of them are now given up.

Near the south end, in excavating for a reservoir, for the Barrow Water-works, there were discovered nine old British cinerary urns, containing human bones; but, strange to say, the workmen destroyed the whole of them, and buried the fragments in the embankment which forms the weir of the reservoir. These urns had been deposited about a yard apart, in a straight line, bearing north-west and south-east, and as there are indications of disturbance of the ground near the place, it is proposed to make a careful search in a little while, and, if more are found — which is very probable — they will have a different fate.

At the village of Ireleth, we enter on a most complicated and difficult district, to correctly interpret which will require a profound knowledge of the science of geology, and as we do not make any pretension to such knowledge, we will tread over it with great diffidence, merely pointing to some of the interesting geological phenomena with which the district abounds. Half a mile east from Ireleth is Haverslack Hill, on the Stewner Bank estate, 604 feet above ordnance high water mark. This hill is principally composed of *amygdaloid*, with a quarry on the summit for getting "road metalling," for which purpose it is well adapted, being tough and hard, abounding with agates, most of the large ones decomposing, — in many instances nothing being left but the matrices in which they were imbedded. Those that are sound take a good polish, and are very beautiful. There is also a large block of green stone lying in the quarry, which, when struck with a hammer, rings like a bell. This *amygdaloid* extends over more than fifty acres — probably over two or three times that area.

One mile south-west of the village of Ireleth, on the Askham estate, we come to the iron ore works of Messrs. Kennedy Brothers. This great deposit of ore is a new discovery, and will reward those spirited and enterprising gentlemen for their many "trial works" in proving new ground. The ore lies in a deep "basin" in the mountain limestone, which here dips at an angle of 40° to the north-west; or, contrary to the general dip of the limestone in Furness. The mines are 100 feet above ordnance high water mark, and distant from the eastern shore of the estuary of Duddon 550 yards, and one and a half miles from the great deposit of iron ore lately discovered at Hodbarrow Point, in Cumberland, which bears nearly north-west from these works.

Although our hæmatite ore is neither in veins, like the copper ore of Coniston, nor stratified like coal, but in " sops " and " basins," one at the end of another, in the mountain limestone; yet, as a general rule, these sops range nearly north-west and south-east. As we have at these works mountain limestone on the surface, only 550 yards from the eastern shore of the estuary, with Hodbarrow Point bearing north-west from them, distant only one and a half miles, it is probable not only that the mountain limestone of Furness is continuous under the estuary of Duddon to Hodbarrow Point, and Limestone Hall, near Sylecroft, but that Dunnerholme is not an outlier, but connected with it, and that there may exist sops and basins of ore under the Duddon, which may be wrought hereafter by means of diagonal shafts, as is done in some of the copper ore mines in Cornwall.

In our traverse from Ireleth village (where we are upon the Lower Ireleth slate) south-west to the iron ore mines last mentioned, where the mountain limestone crops out on the surface, it is evident that we have crossed the junction between these two formations without knowing the exact place; it is also evident that we step from one stratum to the other, without any intervening, so that here the regular sequence of stratification is broken, and the Upper Ireleth slate, the Lower Ludlow rock, and the old red sandstone are wanting. If instead of going south-west we had proceeded due south, opposite to these mines, but on the High Haume estate, we should have come upon a strange mixture of igneous rocks, consisting of *amygdaloid*, green stone, porphyry, with other trap rocks, jumbled together in a strange manner, and beside these some of the deepest stratification of Furness are, by " some great convulsion," brought to the surface, protruded through all the superincumbent stratifications in the series, and repeated in the same order as before. In the northern part of High Haume estate, occur the green slate and porphyry, the Coniston limestone, then a porphyry dyke, and near the east side of the estate, Coniston flag, with a small patch of old red sandstone. On the southern part of the estate, the igneous rocks come in actual contact with the great mountain limestone deposit of Furness. The whole of High Haume estate, as well as the hill on which it is situated, bear evidence of violent mechanical disturbance, and from the great area affected, the centre of it must have been at a considerable depth in the earth. There is evidence also of a local disturbance, which has extended over a very small area, and thrown up two or three cones, or "sugar-loaves," some distance apart, whose bases do not cover more than one acre each, they have had an independent centre of disturbance, and that at no great depth. The Coniston limestone of High Haume abounds with characteristic fossils, but most of them are mutilated, the stone being almost entirely decomposed in the trial quarry where they are found. There is another on the estate, from which a considerable quantity of Coniston limestone was carted to Broughton Mills, for the cement works of Messrs. Rawlinson and Shaw. These works are now given up, and the quarry entirely filled. No organic remains have been found in this quarry, the stone being sound does not show them, although it may contain a great number. If the stone were built into fence walls, and exposed to the atmosphere for a few years, many would be developed. At high Haume, we leave entirely the clay-slate formation of Furness, and no more is seen in a southerly direction, either in Furness or any where else, until we find their equivalents repeated in North Wales (see Sedgwick's letters to Wordsworth). From High Haume eastward to Plumpton Hall — on the shore of Morecambe Bay — we have the junction of the clay-slate and the mountain limestone, and although this is the general direction, it is by no means a straight line, for the two formations are Vandyked into each other. We have given but a very imperfect notice of this strange hill, and we do not offer any other theory to account for its formation than what is self-evident. There are here, known evidence of violent mechanical disturbance, a strange mixture of igneous rocks, and, as the farmer of the estate said, "we have every sort of stone that God Almighty ever made, lying on the hill at High Haume." There is also proof of this disturbance having brought the lowest stratification of Furness to the surface, but not presuming to speculate on the *cause*, we merely point to *some* of its effects. This hill is a favourable site for a view of the surrounding district, and before we descend let us examine some of its physical features. Before us is the real California of Britain — the great hæmatite iron ore district of Furness — some of the principal mines forming a semicircle at the base of the hill. Commencing our review at the north-west, we have first, the new mines of Messrs. Kennedy Brothers, on the Askham estate; almost adjoining these the splendid mines of Messrs. Schneider, Hannay, and Co.,* at Park; close to which are the Ronhead mines of Kennedy Brothers. Glancing round the foot of the hill we find Elliscales, Messrs. Ashburner and Son; Butts Beck, Ricket Hills, Cross Gates, J. Rawlinson, Esq.; the Ure Pits, Ulverston Mining Company; Mousell Mines, Messrs. Schneider, Hannay, and Co.;* all skirting the base of High Haume, and constituting a circular arc of 180 degrees, with a radius of half a mile. Having found so much material of igneous origin at High Haume and Haverslack Hill, and all these extensive mines being so near, we have been very particular in our description as it will naturally form an element in

* This firm has now become the Barrow Hæmatite Iron and Steel and Mining Co.

considering the question of the *formation* of our hæmatite iron ore. Besides the above, there are the mines of Schneider, Hannay, & Co. (the Barrow Hæmatite Iron and Steel and Mining Co.) at Old Hills, Whitriggs, and Marton, all within half a mile of the base of Haverslack Hill, which, as stated before, is principally composed of *amygdaloid*. The circumstance of so many of our mines almost surrounding those two great centres of heat, would naturally suggest the idea that our iron ore is also of igneous origin, but we have for many years been collecting facts bearing on the subject, nearly all of which point a different way. These facts, and the result of our observations, are at the service of any of our geological professors, who may be inclined to grapple with this mysterious question. There are also the mines at Carrkettle, of J. Rawlinson, and the splendid mines of Lindal Moor, Whinfield, Gillbrow, and Whitriggs Bottom, of Messrs. Harrison, Ainslie, and Co.; and at Lindal Cote, those of the Ulverston Mining Company; also at Dalton, those of J. Denney and Co., and J. Rawlinson; and near Highfield House, those of J. Clegg, Esq. All these mining works are within a mile and a half of High Haume; besides, about a mile further, we have the Ulverston Mining Co's. works at Stainton, Bolton Heads, Stone Close, and California; also Mr. Wadham's at Crooklands; and Messrs. Schneider, Hannay, and Co's., (the Barrow Hæmatite Iron and Steel and Mining Co.) near Newton. Perhaps we may have omitted some, but we shall return to this subject.

After leaving High Haume, and journeying southwards to St. Helen's, there is a quarry which produces several species of mountain limestone fossils, the most abundant of which is the *Nematophyllum arachnoidum*, *N. minus*.

A quarter of a mile further south, in Hag Spring Wood, we have the junction of the mountain limestone and new red sandstone, — with this exception there is no well defined boundary between those two formations. The south-western boundary of the mountain limestone deposit seems to begin near Moor Foot, taking a south-easterly direction by Hag Spring, Little Mill, the north side of Furness Abbey, and Park House, from this point, it cannot very closely be traced, there being very slight indications to be seen on the surface, though, without doubt, it is continuous eastward by Gleaston and Gleaston Park to the shore of Morecambe Bay, somewhere between Newbiggin and Moat Hill.

At Hag Spring, near Millwood, the residence of E. Wadham, Esq., the new red sandstone and mountain limestone are on the surface, very near to each other, and we know that the boundary passes on the north side of and not far from Abbots Wood, the seat of J. Ramsden, Esq. After this, in its farther range south-east, it has no defined boundary until it comes to Gleaston, — thence by Gleaston Park to the shore of Morecambe Bay, as before stated.

At Furness Abbey the new red sandstone is overlaid by an immense deposit of diluvial drift, which ranges eastward for about a mile. An excavation for railway purposes has been made on the side of a steep hill of this drift, and in the cutting there is a most interesting section showing the new red sandstone rock, which rises four or five feet from the floor of the excavation, covered by about fifty feet of drift, principally composed of gravel and sandy material, perfectly stratified, containing several large boulders of mountain limestone, also a few large erratic boulders, some of which are of igneous origin. The substances forming the drift include no cementing principle, therefore particles are continually falling, to form eventually a talus at the bottom, and cover up the rock in course of time. This diluvial drift offers some peculiarities which will be noticed hereafter. From Furness Abbey we trace the new red sandstone to Roose, where all rock disappears from the surface entirely in Low Furness.

At Sowerby Hall, one mile from Millwood, there are two or three private quarries in the new red sandstone, but none in work at present, while at Hawcote, half a mile further south, there is a public quarry in active operation, as it has been for generations, producing a good stone, free from iron-bands and other hard veins, and although rather soft, is very durable.

None of the quarries in the Furness new red sandstone have produced organic remains, the stone contains many smooth almond-shaped concretions, but we cannot discover any signs of structure about them.

We have persisted hitherto in calling the sandstone of Furness "new red sandstone," and we are reluctant even now to use any other name, but the progress of discovery compels us to call it "Upper Permian," from Perm, a province in Russia, where the same formation covers more than a thousand square miles of the province.

In a valuable memoir by Sir Roderick Murchison and Professor Harkness, published in the Transactions of the Royal Geological Society of London, very clear reasons are given (unnecessary to detail here), to show that the exact place of our sandstone in the series of stratification was considerably lower than had been given by former observers, therefore it became necessary to change its name, so that in all our future notices we must describe it as "Upper Permian" sandstone.

Half a mile to the south of Hawcote, (in land belonging to His Grace the Duke of Devonshire,) near Ormsgill House, there are two artesian wells; one having a copious flow of very cold and good water, rising two or three inches above the ground, the other not so powerful. The rise of both, summer and winter alike, is constant.

These wells are "bore holes" four or five inches in

diameter, made by Messrs. Schneider, Hannay, & Co., for mining purposes. The water will be highly valuable for the flourishing town of Barrow, (which is not well supplied with water at present,) where it can be conveyed at a light expense, as there is sufficient fall for the water to flow to Barrow by its own gravity.

These wells are interesting as a geological study, for it appears to us that some of the conditions generally considered necessary for the formation of artesian wells are wanting.

A little way south from this point is Ormsgill Nook, opposite to which two tides meet, one flowing from the estuary of Duddon round the north-west end of Walney, through Scarth-hole—the other, from Morecambe Bay, round the south-east end of Walney, up Piel channel, through Barrow and Hindpool, and meeting at this place, they shut off all communication between the main land of Furness and the Isle of Walney.

Half a mile south-east from Ormsgill Nook, are the splendid ironworks of Messrs. Schneider, Hannay, & Co. (the Barrow Hæmatite Iron and Steel and Mining Co.) at Barrow. These enterprising gentlemen, powerfully seconded by the genius and perseverance of J. Ramsden, Esq., have raised, in the course of a few years, the insignificant hamlet of Barrow, into a town of first-rate commercial importance, and its increase and development are so rapid, that streets are rising up on every side, all the houses in them letting to tenants before the foundations are laid, and land in some parts of the town, a few years since valued at £60, is now worth £5000 per acre. The new and extensive steel works erected at Barrow, will really be a principal attraction to strangers visiting the town. However, we will not attempt a description of these public works, (which would require a volume to do them justice,) but leave it to abler hands.

We now cross over to the north end of the Isle of Walney, the extreme point of which is the great breeding ground for several species of sea fowl, viz., Shell Duck *(Tadorna vulpanser)*, Ring Plover *(Charadrius hiaticula)*, Dunlin *(Pelidna variabilis)*, Oyster Catcher *(Hæmatopus ostralegus)*, Common Tern *(Sterna hirundo)*, Sandwich Tern (*Sterna cantiaca*), Lesser Tern (*Sterna minuta*), Black-backed Gull (*Larus marinus*), Common Gull (*Larus canus*), Wild Duck (*Anas boschas*), Lesser Black-backed Gull (*Larus fuscus*).

Having for many years devoted a considerable portion of time in studying the manners and habits of the wild animals and birds of our district, both in the field, on the mountain, and by the sea-shore, the result has been that our observations have imperceptibly accumulated to a considerable extent, and from them we extract the following anecdotes respecting the Shell Duck.

No doubt our friends, the naturalists, will have seen the *ruse* practised by the *Pulvatores*, viz., the partridge, grouse, black-cock, &c. On approaching their nests, particularly when they have a young brood only a few days old, the old birds will simulate every kind of injury, such as a broken wing, a broken leg, &c. They will run and tumble before you, with their feathers ruffled, and seemingly not able to fly a yard, so that you may catch them in a moment, but no sooner have they succeeded in drawing you away from their nests, than up they get, fine, plump birds, and fly off exulting in the stratagem. Certainly this is no bad manœuvre, but the Shell Duck displays much greater vigilance to prevent surprise, and superior artifice to escape from danger, than any practised by the *Pulvatores*, as the following account will show.

One fine, warm, summer day, we saw on a dry sand-bank, at the north-east end of Walney, about two or three hundred yards from the shore, three shell ducks, quite still, enjoying the sunshine, two of them with their heads under their wings, the third wide awake, keeping guard while the others slept. We laid down behind a sand-hill and watched them a long time. At length the second duck put up its head, the first taking a turn for repose, and after about the same lapse of time, the third duck looked up, while the first and second slept, and thus they relieved each other like soldiers on sentry. When we showed a little above the sand-hill, and made an attempt to creep a little nearer, all their heads were up in an instant, the sentinel duck had given the signal, which we could not hear, though we listened attentively. After we had lain down for some time, two of them put their heads under their wings again, the third keeping watch as before, and in this way they changed guard, alternately watching and resting, until we stood up, and came out on the sea-shore, when they all took flight, showing great vigilance and acuteness to prevent the approach of danger.

We then proceeded round the end of the island, and in turning a sharp point at the Scarth-hole, we came upon a shell duck with a young brood of very small ones, about as large as sparrows, not more than three or four days old, and quite unable to fly. The old lady duck (Mrs. Tadorna) flew off instantly, and left her young brood to take care of themselves, which she well knew they were able to do effectively, as she had given them a lesson which showed a degree of sagacity and cunning far superior to any practised by the *Pulvatores* when an enemy was nigh. It was as follows:—As soon as the old duck had left her young, they began to swim round, in a small circle, three or four times, until the whole of them went down suddenly with a sort of jerk, throwing the water about, and making a considerable disturbance in it, remaining below three or four seconds, when they came to the surface one less in number. They swam as before, enlarging the circle every time, dipping after awhile again, with a sharp jerk, and when they rose one more was missing, and thus they continued alternately swimming round and diving, one of them disappearing at a time, until the whole had vanished beneath the water. Knowing that no species of the duck family is amphibious, and as none of the ducklings could possibly leave the Scarth-hole without our seeing, it was evident they were just below the surface, protruding a portion of their bills above the water, so as to enable them to breathe, and in this situation they would remain until we had left the place.

During the breeding season, the ground is almost literally covered with eggs, and later on young birds are running about your feet, covered with a woolly down, looking like so many lambs in miniature.

The sea-shore abounds with recent shells, of several species. There may also be found in the shallow pools on the scars, the beautiful Sea-anemone (*Actinia mesembryanthemum*), with its minute waving tentacules, also the Hermit Crab *(Pagurus Bernardus)*, and many other curious forms of animal life, some of which will be noticed hereafter.

We are not aware that rock of any kind has been found in Walney, except the "crab rock," which in reality is not a rock, but a soft, recent living coral, in which the crab-fish excavate and harbour, where they are industriously hunted afterwards by the fishwomen, at low water, during the months of April, May, June, and July, at which time they are in proper season. Lobsters are occasionally found in these burrows in the crab-rock, and also the cuttle fish (*Sepia officinalis*). There is also a bed of oysters, which can only be reached at the low water of a twenty feet tide.

In traversing the dreary south-east side of the Isle of

Walney, we shall probably find at high water mark, several species of star-fish, *spatangus*, &c., &c., which have been thrown by storms high up the beach, where the tide will not reach them again for several weeks, and when all the animal matter is decomposed, and the bones bleached by the atmosphere, they are perfectly cured and free from smell, forming beautiful specimens for the cabinet.

Walney affords a geological lesson of some interest, as illustrating the powerful agency of water (when agitated by strong currents or storms) in transporting heavy bodies to great distances. We may state it as follows. Although there is no rock on Walney, the greater part of the sea-shore, all round the island, is covered with shingle, and it is somewhat singular, that nearly the whole of it is composed of granite, and hundreds of cart-loads of boulders have been used for building purposes, both in dwelling houses and fence walls. These boulder walls are very good and durable when laid with mortar.

As all the rock formations of Low Furness (the nearest land to Walney) are composed entirely either of mountain limestone or upper Permian sandstone, and as the shingle on the sea-shore is principally made up of the *debris* of these rocks, it is remarkable that so little of this material should be found in Walney, especially as all the larger boulders are of granite, some of them weighing several hundred pounds, all of which must have been transported at least twenty miles from the parent rock in Eskdale. This fact does not admit of doubt, as there is no other granite rock in *situ* from which it could have been conveyed.

The south-east side of the island is notoriously dangerous. The "back of Waana" has ever been a terror to seamen, and will remain so through all time. Many a gallant ship has gone to pieces there, and it is supposed that one of the ships of the Spanish Armada was wrecked on this coast. There is some probability that such was the fact, as several pieces of ancient ordnance have been found at different times by the inhabitants of Biggar, about two miles from the south-east end of the island, and converted into implements of husbandry, most of them being composed of malleable iron. This circumstance becoming known to C. D. Archibald, Esq., about the year 1838, he caused excavations to be made at the place, and discovered several similar pieces, which we believe are still in his possession.

Opposite to Biggar, and about a mile south from it, are the remains of a forest of timber trees, the roots still fixed in the ground. These can only be seen when there is a "good ebb," viz., the ebb from a twenty to a twenty-one feet tide.

Journeying from Biggar to the "South-end," or rather the south-east end of Walney, a distance of two miles, we arrive at the lighthouse, our first starting point, having completed the whole circuit of Furness, and noticed slightly the different objects of interest, as we passed them, many of which, although not strictly geological, have some bearing on the science.

Before leaving Walney, we may notice a curious geological phenomenon — the lengthening of the island, at its extreme south-east point, by immense waves of shingle, thrown up by storm — these shingle waves resemble the swathes of grass in a newly-mown hay field, but instead of being straight lines, they are rather quick curves, and truly concentric.

We shall now endeavour to describe the stratifications of Furness, where they may be observed in *situ*, with the different species of Organic Remains characteristic of each formation, beginning with the lowest, *i. e.* the oldest, and rising in the series until we reach the highest, viz., the Upper Permian Sandstone of Hawcote and Furness Abbey.

STRATIFICATIONS OF FURNESS.

GREEN SLATE AND PORPHYRY.

As before stated, if a line be drawn from Duddon Bridge to Brathay Bridge, all that part of Furness on the northwest of it, (which is also the extreme north-west portion of the County of Lancaster), is due to the Green Slate and Porphyry formation, including Ulpha, Seathwaite, Tilberthwaite, Monk-Coniston, Elterwater, and Dunnerdale, comprising an area of about forty square miles. No organic remains have yet been discovered in this formation, but as the Skiddaw slate — the next stratification in the series of deposit *below* the Green Slate and Porphyry — has been proved to contain several species of graptolites and other Cambrian fossils, we may reasonably expect that fossils will be discovered in the Green Slate and Porphyry also, when it has been thoroughly examined.

Walney Scar quarries, in Seathwaite, produce some very beautiful arborescent slates, representing woods, groves, &c., in many instances single trees are represented with the greatest possible clearness. These beautiful figures are not organic, but merely superficial markings due to manganese, and in every instance of their occurrence, the slate rock contains minute cracks, admitting moisture to the interior of the stone, therefore, if an attempt were made to polish them after the manner of marble, the markings would be entirely obliterated. However, these elegant objects can be brought out perfectly, and made permanent, by giving them a slight coating of Venice turpentine, with a camel's hair pencil. Besides these, there are splendid specimens of slate, illustrating faults, throws, waves, contortions, &c., equal, if not superior, to the same phenomena at Common-wood quarries, on the Cumberland side of the river Duddon, in Ulpha. The quarries of Walney Scar are especially instructive, as furnishing data bearing on the question of "slaty cleavage," which is by no means well understood at the present day. Various theories have been propounded to account for it, all of them more or less beset with difficulties, but certain isolated facts are known respecting it, while the main question of "slaty cleavage" still remains somewhat of a mystery.

The principal quarry contains considerable masses of Brixiated rock, composed of large and small angular fragments of various other rocks, very closely and firmly cemented together with Felspathic matter. This Brixiated mass has a cleavage almost as perfect as the pure slate material itself, and when split up the fracture passes through the whole without leaving any of its component parts in relief, which is the case also with all the waved and contorted masses abounding in these quarries — thus demonstrating that "slaty cleavage" was induced after the rock had become perfectly hard and consolidated. In this slate, the plain of cleavage forms a high angle with the plain of deposition, the different colours in their striping being traceable on the face of the rock, and lines of deposit no thicker than paper are defined with the utmost clearness. Some of this faulted and contorted slate is now used for mantel pieces, table tops, and various ornamental purposes. There are other matters of interest in these quarries hitherto unnoticed, and we believe that a minute examination of them would be an important and valuable geological lesson, more calculated to impress on the mind the different phenomena there presented, than a considerable amount of reading at home. Besides these interesting slate works on Walney Scar, there are several others in Tilberthwaite, near Elterwater, and on the "Old Man" mountain in Coniston. All the mountains of Furness are Green Slate and Porphyry, viz., Coniston Old Man, Grey Friars, Dowcrags, Lambcrags, Wrynose, Leather-barrow, &c.

The copper ore mines of Greenburn, Seathwaite, Cockley Bank, and the splendid mines of Coniston, are all in the Green Slate and Porphyry formation, as are also the iron ore mines of Dunnerdale. The Green Slate of this district is very hard and durable, even more so than the Upper Ireleth slate, although it does not give so smooth a face. It is a good building stone, but has not been extensively used for that purpose, the district where it is obtained being rather thinly inhabited, the principal use for it is roofing slate. It does not decompose rapidly like the Skiddaw Slate, by atmospheric influence, so that it presents many jagged and serrated steeps, particularly at Dowcrags, Lambcrags, and near Greenburn Mines. At Dowcrags sometimes may be seen a spectacle of no common interest, the immense cliffs for a considerable distance are perpendicular, and in some places overhang the base, containing fissures and joints which admit moisture, and in some parts water may be seen trickling down the rock. A hard frost in winter converts all this into ice, which by its natural expansion forces off from the parent rock large masses of stone, to bound and crash along the steep descent at the foot of the cliffs with a force which nothing can resist, some large blocks even entering Goatswater before their progress is arrested, the border of that small lake being covered with their *debris* for many acres in extent, several single stones lying there weighing about three or four hundred tons each.

This immense accumulation of loose stones is known as the "Fox Bields," indeed it seems all the foxes of the Coniston district look upon it as a refuge, because when pursued by hunters they make for it, and if they can reach the place safely they are secure from further pursuit.

The Green Slate and Porphyry presents some curious examples of jointing, viz., dip-joints, strike-joints, and in some instances crossings in a diagonal direction, forming the whole face of the rock into a series of triangles. A singular instance of jointing, of a different kind, occurs on a floor of rock, to be seen on the east side of the road between Mr. Barratt's house and the great copper mines at Coniston, the whole face of the rock, at a little distance, having the appearance of an immense fossil coral, somewhat resembling the *Nematophyllum arachnoidum*, the several stars of course larger, being about four inches in diameter.

The Copper Ore Mines of Coniston are very instructive, and well deserve a visit from every mineralogist in the kingdom. We will not attempt to describe them, as an extract from C. M. Jopling's "Furness and Cartmel," will serve our purpose better. "The copper ore mines now worked at Coniston, are situated a little way up the base of the "Old Man" mountain. These works are exceedingly interesting, and afford a complete contrast to the simple mode of mining for iron ; the copper ore, after being procured, undergoing a long preparatory process. The entrance to these mines is by means of an adil or level (a small tunnel) driven into the side of the hill. On entering, daylight soon fails, and the flickering candle, stuck into a lump of clay which serves as a candlestick, is just sufficient to light the way without dissipating the gloom. The hollow sound of the distant blast rumbles through the mine; whilst the waggons rattle along, making the long galleries echo with an almost deafening noise. After proceeding some distance, the sharp blows of the miners' picks are heard, and the men are soon discovered pursuing the yellow metal, which gleams brightly by the candle's light. Boring operations for blasting are also going on. The quick movements of the miners, the depth of shade, and the partial illumination, form a highly picturesque scene. In some parts of the vein the ore has been worked out to a great height, and the roof is not visible in the thick darkness which prevails. From this tier of workings the descent is made, by means of shafts, to others many fathoms lower.

"The scene on the exterior is no less interesting. The ore, after having been brought out, is broken into smaller pieces, first by means of hammers, and then by stamps, put in motion by a water-wheel; it is then picked over by a number of children, who separate the pure pieces of ore from the impure; the latter are again broken, and undergo the process termed "jigging," in which operation the metal, being heavier than the sparry matter with which it is associated, sinks to the bottom, and the refuse is taken off from the top. There are other plans adopted, by which a still further separation is effected of any particles of ore that may be left among the refuse.

" The copper veins here, it is stated, are generally marked by water-courses. In the veins there occur, beside the copper, lead ore rich in silver, black jack, numdic or sulphuret of iron, and small quantities of foliated native copper. Some very beautiful specimens of the peacock copper ore are found, and a little of the green carbonate. The principal ore, however, is the yellow sulphuret."

CONISTON LIMESTONE.

The next stratification above the Green Slate and Porphyry is the Coniston Limestone. This formation, although not valuable for economical purposes, is of the very highest geological interest, and was taken by Professor Sedgwick for the base of his Geological Section of Furness, (see Sedgwick's Geological Letters to Wordsworth, published by Hudson of Kendal,) being exceedingly well adapted for the purpose, as it ranges entirely across the Furness district, extending many miles into Westmorland, and in the contrary direction into Cumberland. As all the different stratifications of Furness run parallel with the Coniston Limestone, it will be necessary to define it clearly to enable us to understand all the other formations, and for this purpose we must consider the geographical figure of the district, and impress it on the memory. If we examine any good map of the County of Lancaster, we shall see that the general figure of Furness is a long oval, twenty-seven miles in length, from north to south, and eleven miles from west to east, *i. e.*, from Duddon Bridge to the middle of Windermere at a point about a mile north of Newby Bridge. This line is due east and west, and would cut Furness into two nearly equal parts, but if instead of going from east to west, as above indicated, we start from Duddon Bridge, and take a north-east direction across Furness, we should end our journey near Brathay Bridge. This line would be fourteen miles in length, and nearly represents the boundary between the Coniston Limestone and the Green Slate and Porphyry, which is also the boundary between the fossiliferous stratifications in the south-east, and the non-fossiliferous Green Slate and Porphyry in the north-west, already described.

As the Coniston Limestone is of such high geological importance, we shall give a somewhat detailed account of it, first showing its range, and afterwards noticing the numerous species of beautiful Organic Remains with which it abounds. This limestone has until lately been considered the lowest formation in the North of England that contained any amount of Organic Remains, but within the last few years it has been discovered that the Skiddaw slate, a lower stratification, is somewhat fossiliferous also, though not to be compared with this formation in amount or variety of species. The Coniston Limestone, as before stated, is a hard, compact, dark blue rock. It is not much used as a building stone, being subject to rapid decay by atmospheric influence, and throughout its whole range we have the Green Slate and Porphyry below, and the Coniston flag above it, both of them good building materials, particularly the latter, neither is it much used for burning into lime, being subject to run in the kiln, almost any foreign matter acting as a flux. A farmer in Low Furness, a few years since, tried an experiment with a few tons, but instead of making agricultural lime, it melted into slag at the bottom of the kiln, becoming a solid substance, incapable of extraction by any means, and the kiln was ultimately taken down.

A considerable quantity of this limestone was, a few years since, manufactured into cement, at Broughton Mills, but these works are given up, and it is now principally used for fence walls, notwithstanding that it is of high geological interest, and known to lovers of the science in all parts of the world.

The Coniston Limestone first appears on the surface at Beck Farm, in Millom, which may be considered its south-west limit, and from this point we propose to follow it on its line of strike to the north-east, for thirty or forty miles, noticing whatever we may deem of sufficient interest in this long traverse. On the Beck Estate, there is a private quarry, where little work has been done for several years past, containing a great many nodules, or concretions, of limestone, most of them ranged in lines in a singular manner. Very few Organic Remains have been obtained, the rock being not much decomposed, but in the adjoining fields, corals of two or three species are turned up with the plough, particularly the *Palæopora megastoma* (Mc.Coy.) The Coniston Limestone here has a thickness of 500 or 600 feet, which is considerably more than its average thickness in other places; but this great development diminishes rapidly as we proceed on our course to the north-east. A little to the west of Millom Castle, this rock crops out again by the road side, very much broken up and decomposed, where we have obtained fragments of several species of Organic Remains, one of which we believe to be new, of the type of the *Edmundi sulcata*, a characteristic fossil of the mountain limestone, but much smaller.

Half a mile north-east of Millom Castle is the farm of Waterblain, where there is another private quarry in the limestone, but as we are near to the iron pyrites mines of "Hill," and two porphyry dykes which cross the public road, a short distance west from the village, they have somewhat changed the natural appearance of the limestone rock.

We will now pass on to Graystone House, a distance of two miles, where the limestone crops out for a considerable distance, and a quarry that has not been in work for several years shows the limestone rock different from that of any other part in the whole range of this formation, "being mineralised by heat from its proximity to the great porphyry dykes which lie on the west side of it" (Sedgwick). This change from the normal condition gives it a very unpleasant harshness to the touch. We have obtained several specimens of Organic Remains at this place, but they were most of them broken and otherwise imperfect.

After leaving Graystone House, we cross the river Duddon a little below Duddon Bridge, and enter Furness, preceeding up the valley of the Lickle, on the east side of Bleansley Bank, to Hartley Ground, where there is a considerable development of Coniston Limestone, but it begins here to thin rapidly, until at Broughton Mills it crosses the public road only nine yards in thickness, and is seen to crop out and form part of the floor of the public-house near the Mills. Proceeding to Appletreeworth we find the limestone again in force,—thus far on our route to the north-east, not many Organic Remains have been found, the principal part of the rock being compact and sound; and, although the latter may abound with fossils, few can be seen until it becomes decomposed, when hundreds will be developed. A portion of the rock, near Appletreeworth, is much disintegrated, and full of fossils of several different species, particularly the *Orthis porcata* (Mc.Coy.)

There is also a mineral vein of a very dark colour, almost black, four or five feet in width, and a drift, or level, driven into the vein to a considerable distance, commencing at the foot of a steep slope, and entering directly into the hill. This driftway has not been made within the memory of any one now living, neither is it known who were the workers. The vein shows some indications of lead ore and of having been subjected to heat. Within a few yards of the entrance a curious soft, fissile, rock crops out, of a beautiful yellow colour, free from gritty particles, which we have used for the purpose of polishing marble, and found it very little, if any, inferior to the "Water of Ayr" stone for that purpose. It rubs down to a very smooth surface, is free from pores, and a few feet below the surface, where not subject to atmospheric influence, is sufficiently compact to be used for lithographic purposes, as it very much resembles the lithographic stone of Solenhopen.

This deposit appears to be just at the junction of the Coniston Limestone, with the Coniston flag, and probably is an altered condition of the latter formation, due to the same cause producing the mineral vein with which it is in contact, and whether such be the case or not it is very singular and deserves a minute examination.

From Appletreeworth the Coniston Limestone continues on the surface, and ranges over Broughton High Common to Ash-ghyll Quarry, in Torver, a distance of three or four miles, forming a long narrow ridge nearly the whole way, having the Green slate and porphyry on one side, and the Coniston flag on the other, rock of both formations being in close proximity to it, and in some places seeming to burst through and rise a considerable height above them, having the appearance of an igneous dyke at a distance, but it is easily distinguishable from other formations by its weathering with a rough and honeycombed crust on the surface.

Between Appletreeworth and Ash-ghyll there is nothing of particular geological interest, but near to Ash-ghyll quarry there is a small hill, which rises abruptly above the general surface of the ground, a portion of the rock and shingle of which is highly decomposed, and abounds with fossils. The Coniston Limestone ranges alongside, and is in actual contact with Ash-ghyll quarry, although it is a slate work in the Coniston flag. This place contains a few species of fossils which will be noticed when we describe the range of the Coniston flag formation.

CONISTON LIMESTONE. 53

We now trace the Coniston Limestone in a north-east direction, across Torver Common, to where it skirts the foot of Coniston Old Man, at which place there is exposed an extensive bed of limestone shingle, and although it may not be the best locality for Organic Remains, yet we have obtained several rare species of fossils from it.

This shingle bed may also be considered the commencement of a splendid geological field, presenting something of interest almost at every step, where thousands upon thousands of Organic Remains may be developed wherever the rock has suffered any amount of decomposition, but where sound and compact, a search is almost useless.

The fossils of the Coniston Limestone are especially interesting, as giving almost the first indications of organic life on the earth, and as they are all of high organization, their markings and sculpturings are beautifully sharp and clear, and in this respect much superior to the Organic Remains belonging to some of the higher and newer formations. Here is afforded a strong argument against the sceptical theory propounded by the author of the "Vestiges of Creation," (that all organisms originated from a monad without structure or organization, and rose step by step to the highest state of organization,) which was so ably refuted by the late Hugh Miller, in his "Astrealepus,"* it is even more forcible and convincing than the argument brought forward by him, inasmuch as the Coniston Limestone is immensely older than the old red sandstone of Scotland, in which the *Astrealepus* is found, and as they exist in the greatest profusion and variety for a distance of two or three miles, it is evidently a geological field of great interest and importance, and deserves a more careful examination than it has hitherto received.

We have visited for a few days this portion of its range

* *Astrealepus*, a fossil fish, of high organization, found in the old red sandstone of Scotland.

several times, and always with increased pleasure, never ceasing from work amongst the rock during rain or fair, to be always rewarded by finding some new species of Organic Remains, or at least more perfect specimens of fossils already known to science, and we invariably left Coniston with great reluctance, believing it to be the best locality in Britain for obtaining specimens from the old rocks. We would advise the geological student who is anxious to obtain a collection of these interesting relics, and to study the Cambrian system of Professor Sedgwick, as developed in the Lake District, to proceed direct to Coniston, and engage private lodgings for a few weeks, (he cannot go wrong, because all are respectable,) provide himself with a Sedgwick's geological steel hammer, two pounds in weight, with a shaft fourteen inches long, to break up large pieces of stone, and a small hammer, four ounces in weight, to dress his specimens with afterwards. An iron chisel, eighteen inches long, tipped with steel, will be found useful to break or wrench off pieces from the rock, where it is somewhat broken and decomposed, for such places will afford him the best specimens, and a wallet 48 inches in length by 18 in breadth, to carry home his specimens, will complete his outfit. A wallet is much better than a bag, as it will have a broad bearing, and not cut the shoulder like the straps of a bag, therefore it will enable him to convey a considerable weight with ease and comfort. A plentiful supply of paper is necessary to fold specimens in, to prevent scratching, which would almost spoil them, and a few chip boxes with sufficient cotton wool to pack and secure small and delicate fossils, so that they will not either shake in the box or touch each other.

We will now suppose the student furnished with all the above requisite appliances, and having brought him step by step from Millom, by Duddon Bridge, Broughton Mills, Appletreeworth, and Ash-ghyll, to this most interesting geological field at Coniston, we will mark out for him a day's work, giving such minute particulars, as will enable him to a certainty to bring home at night a wallet load of these beautiful Organic Remains.

Taking it for granted that he is a stranger, or at least a person not well acquainted with the different localities around Coniston, our directions are as follows: — As soon as breakfast is over, shoulder your wallet containing all your implements for work, make direct for Waterhead House, the seat of James Garth Marshall, Esq., and enter his park by a gate on the north side of the public road, nearly opposite the house.* Follow the private road leading through the park, in a north-east direction, into an adjoining wood, to be continued into some rocky pasture fields, where you may spend not only this day, but weeks and months, without being able to exhaust the rocks of all their treasures. Before leaving the wood you will find a part of the road cut through the limestone rock and a small quantity of shingle lying by the road side, in which you will probably find the *nebulipora lens* (Mc.Coy), *Favocites crassa* (Mc.Coy), *Rhynchonella tripartita* (Sowerby), and others, thence pass through the wood and the first rough field, you will easily distinguish the Coniston Limestone by its colour from the Green slate and Porphyry on the north side of it, with which it forms a junction in the same field. As you proceed you will find other beds of shingle, all of them good for Organic Remains, but look out for broken and decomposed rock which you will find to a certainty in a little while, as there are several examples in this field, so you may take your choice, but the best

* You need not be afraid, for if you meet the gamekeeper, who is a very civil man, he will not molest you, knowing that no one goes poaching with a hammer and chisel in his hand and such implements as your wallet contains; on the contrary, he will guess your errand without your telling it, and if you should fall in with Mr. Marshall, so much the better, for he is an accomplished geologist himself, and has read before the British Association some highly valuable papers, illustrated by beautiful diagrams, of the geology of the Lake District.

is a large rock standing boldly up above the ground, with the north side greatly shattered. This rock abounds with numerous species of Organic Remains, of several different genera, indeed it is inexhaustible, so make it the limit of your day's journey and begin work at once, in the following manner: — First empty the wallet and spread it on the ground to make a comfortable seat near your work, to be commenced by inserting your chisel into one of the numerous cracks or fissures in the rock, and in this way, without exerting much power, large pieces can be wrenched off, which must be carefully laid down near. Continue in this manner until you have obtained a good number, then take your large hammer and begin to split up all the pieces, first examining them carefully on both sides, for fear you may spoil some good fossil. A considerable portion of this rock will split in the plane of deposition, and when one of these pieces is opened by the hammer, it sometimes developes a face entirely covered with beautiful Organic Remains, which should be carefully laid on one side, until you have finished for the day. Some of them will split two or three times, therefore, if there be nothing good on the first face, break it open again, the next may be better, or the following. This is something like school-boys playing at "picture over leaf" with Æsop's Fables, but far more interesting, inasmuch as the different pictures in the Fables are all known, while these pictures have not been seen by any eye, and frequently a fossil is developed in this manner of which the species or even the genus is unknown. When all have been broken up and examined, and the best laid carefully on one side, without touching each other, resume the work of the chisel for awhile, and continue in this manner, alternately breaking from the rock and splitting up with the hammers, and in five or six hours you will be entirely surrounded with specimens, all of them more or less beautiful. Probably smitten with the geological malady, your mind will be so much absorbed with the contemplation of the wonderful organisms by which you are surrounded, that, judging by ourselves, it will act upon you with a sort of mesmeric influence, and you will ponder and dream over these strange but highly organized forms until you forget (for a while) everything else in the world. This is truly " the fossil poetry of the earth, and such a poetry as those can never dream of who in a pebble see a pebble and nothing more."

You will now begin to prepare for returning home, by carefully folding each specimen in a paper, and it will be advantageous to place a soft paper on the face of tender fossils, which may afterwards be wrapped in a stronger cover. Pack your wallet as cautiously as if the specimens were eggs, and do not disturb or jolt them when on your shoulder, or they will be injured by the friction; but if you can take them home without setting down or changing shoulders, they will be as perfect as when first folded up. It is probable you will find some too tender to be carried in the wallet without breaking, and these must be packed separately in the chip boxes, and made secure with cotton wool to prevent shaking, so as to go safely in your pockets. After refreshment at your lodgings, unfold your specimens, examine and label all that you know [*] and be particular to give the locality where found, as this is important in all cases. Notwithstanding all your care, there will be many you cannot name, some of which may be new, and others, although not absolutely new, may not have been found before.

You will probably have discovered in this day's work, most of the species named in the following list, viz.,— *Orthis porcata* (Mc.Coy), *Palæopera megastoma* (Mc.Coy), *Favocites crassa* (Mc.Coy), *Nebulipora explanata*, *Berenicea*

[*] In this you will be assisted by Professor Sedgwick's splendid work "Synopsis of the Classification of the British Palæzoic Rocks." Deighton, Bell, & Co., Cambridge, 1855.

heturogyra, Retepora hisingera, Lichas subpropinqua, Zethus atractopyge, Z. rugosa, Calymene subdiadimata, Spirifera percrassa, Orthis porcata, Lingula Davisii, L. ovata, Petraia, æquisulcata, Ptiladicta explanata (Mc.Coy), *Tentaculites anulatus* (Schlot), *Chasmops Odina*, (Eichw. S. P.), *Orthis vespertilio* (Sowerby).

Besides the places described above there are several others where Organic Remains may be obtained in abundance, indeed wherever the limestone rock presents the same appearance as the last, and we would advise you to look for such spots, as you will obtain different species in each locality.

We will bid good-bye to the student for the present, as he will be anxious to go to the same rocks again in the morning, and many succeeding days, and after this long digression we must return to the foot of Coniston Old Man, thence through the fields by the "Scrow," and down the deep descent to Coniston. At Coniston the limestone may be seen crossing the Town Beck, behind the Black Bull Inn, where it is a fine dark blue rock, the lawn in front of Mr. Barratt's house, again by the Rev. T. Tolming's, and through the village till it skirts the foot of Lambcrags.‡

From Sunny Brow forwards for two or three miles there is nothing of particular geological interest, but near Pullwyke a considerable quantity of rock was quarried a few years since, to build fence walls inclosing some extensive gardens. We examined this stone carefully at the time of its excavation; it was then a fine sound dark limestone rock, and had all the appearance of being non-fossiliferous, but some of it is now beginning to decompose, and it

‡ Lambcrags is a stupendous natural rockery of the Green Slate and Porphyry formation, several hundred feet in height, and when seen from the interior of Mr. Marshall's house, at Waterhead, reflected by a mirror in one of the rooms opposite, it has a truly magical effect.

turns out, as we expected, to be highly fossiliferous, thus proving the truth of what we before advanced, that where this rock is perfectly sound and strong, scarcely any trace of Organic Remains can be found, although at the same time it abounds with them, but when built into fence walls and exposed to storms for a few years, it begins to crumble and show Organic Remains in thousands. A still more remarkable example of this phenomenon may be seen on Applethwaite Common, † in Westmorland, at which place the Coniston Limestone rock *in situ* is good, and Organic Remains are very rare indeed, but the walls on each side of the public road over the common were built with this stone some thirty or forty years ago, and are now so much decomposed that they are almost a heap of rubbish, yet they exhibit innumerable forms, and offer a good field for the student. This principle holds goods to some extent also with reference to the Lower and Upper Ireleth slate.

We have now traced the Coniston Limestone from Beck, in Millom, (where it first shows on the surface), step by step across the whole of Furness, in a north-east direction, noticing the different points of interest presented in our traverse, but we are now about to lose it in Furness altogether, for at a short distance from Pull-wyke, the stratification enters Windermere Lake, and issues from it half a mile north of Low Wood Inn, in Westmorland, thence coursing up the rough ground to Troutbeck, over Applethwaite Common through Kentmere, and over the next ridge of hills into Longsleddale.

Thus far we have followed it on the line of strike, and according to Professor Sedgwick, who is a most accurate observer, it then ranges by Brotherdale Head, to near

† Applethwaite Common is an extensive plot of common land, situate on both sides of the public road leading from Troutbeck to Kentmere, and forms a portion of the mountain ridge of which "Hill Bell" is the highest point.

Shap Wells, when it is lost under the peat bogs, but this indefatigable geologist has found it again in Dent, and has there obtained from it many of the characteristic fossils of the Coniston Limestone, thus showing that this interesting and geologically important formation has been acted on by some powerful agent which has dislocated and thrown it some miles to the south. The Coniston Limestone is a long narrow band upwards of fifty miles in length, and for a great part of that distance may be seen above ground, yet the breadth it occupies on the surface, is so trifling, that if it were laid down to a scale on a geological map of England, it would be truly represented by a fine line drawn by the pen which would have a small crook or notch in it at Coniston, and a considerable break between Shap Well and Dent.

We will leave it for the present, and return to the neighbourhood of Duddon Bridge, to commence a second traverse over Furness on the Coniston Flag, that being the next formation above the Coniston Limestone in the series of stratification.

CONISTON FLAG.

The foregoing chapter has been devoted entirely to a description of the Coniston limestone, giving in detail its geographical range, and noticing some of the beautiful Organic Remains with which it abounds. The next in the series of stratification is the Coniston Flag, a highly valuable stone, being used extensively for building, and for flags and roofing slate, for all of which purposes no better material can be found in England.

CONISTON FLAG.

This formation stretches entirely across Furness, parallel, and in immediate contact with the Coniston limestone, from the estuary of Duddon, near Broughton, to Windermere Lake, near Brathay, and it may be seen on the surface the whole distance between these two points. In some places it is moderately, but in no part of its range abundantly fossiliferous, it yields a very good building stone throughout its course, and in some places, which will be particularly noticed as we reach them, very superior indeed.

The Coniston Flag, like the Coniston limestone, first shows itself in Cumberland, near Holborn Hill, Millom, then ranges north-east, skirting the low lands, by which it is partially covered on the east side of Underhill and Green-road Railway Stations, thence forwards towards Duddon Bridge. After crossing the Duddon it is again concealed for a little way, but it soon emerges and shows boldly before we arrive at Broughton, and so continues to Windermere Lake.

The breadth which the Coniston Flag occupies on the surface in the neighbourhood of Broughton, is at least a mile and a half, viz., from the valley of the Lickle on the north-west, to near Foxfield Railway Station on the south-east, where it forms a junction with the Coniston grit.

The railway, the whole distance from Broughton to Coniston, is made almost entirely in the Coniston Flag, and it exposes some instructive sections illustrating dip joints, strike joints, &c., and in one of the cuttings between Woodland and Torver the face of the rock is almost completely divided into lozenge shaped pieces a few inches in diameter. These rock cuttings also show various other natural phenomena. The student must not neglect to examine an interesting section of rock exposed by a recent excavation immediately behind the Railway Station at Coniston. In this long line from the estuary of the Duddon, near Duddon Bridge, to Torver and Coniston,

Organic Remains are exceedingly rare, very few having been found in any part of it. We will now traverse the north-east boundary of the Coniston Flag, commencing again near Duddon Bridge, and proceeding up the valley of the River Lickle towards Broughton Mills. This formation is partly covered for a little distance, but it is well developed in Broughton Tower Park, and near Hawthwaite, from thence by Lane-end, Broughton Mills, and Hopkin Ground, some trifling excavations have been made at two or three different places. Between Hopkin Ground and Appletreeworth, Organic Remains of five or six different species will be found on the fence walls.

Beyond Appletreeworth, on Broughton High Common, we come to Mr. Postlethwaite's slate quarry in the Coniston Flag formation, which is in active work and produces a quantity of very good roofing slate. Near the north-east boundary of Broughton High Common we come to another quarry of slate, belonging to and wrought by the Lord of the Manor, John Sawrey, Esq., of Broughton Tower, producing excellent roofing material, but it is difficult to work, being situate in the bottom of a valley, and the whole of the slate metal has to be raised by mechanical power. We now enter on Torver Common, the royalty of which belongs to the Crown, and we pass several trial quarries where quantities of *debris* remain on the surface, all of which produce Organic Remains, more or less, but by no means abundant in any of them. The next principal quarry is Ash-ghyll, in Torver, also a slate quarry in the Coniston Flag formation, abutting upon, and in actual contact with the Coniston limestone. It contains several species of Organic Remains, some of which are of the type of the Coniston limestone, but as the rock is perfectly sound in every part, the fossils cannot be clearly developed; notwithstanding this it may be considered one of the best places for obtaining them in quantity.

The following species have been found, viz.:—*Spirifer biforcata*, Var. *fissicostata*, *Orthis protensa*, *Orthoceras sub-undulatum*, *Odontochile obtusa caudata*, *Orthis filosum*, *O. laqueatum*,—*tenuiciactum*, *Cycloceras sub-anulatum*. Besides the above there are one or two others which seem to be peculiar to this place, and in the trial quarries near, two or three species of graptolites may be obtained, but their markings are not distinct.

Ash-ghyll is an old quarry, which has formerly been worked by the dangerous and imperfect method of galleries and tunnels, and as a proof of the great danger of this system of working we may state that in one of our visits to this place, accompanied by a friend, we spent some hours geologising in the large chambers and tunnels, but on our next visit a little while after, the immense superincumbent mass above had crushed in the whole chambers, several thousand tons of stone having fallen to the bottom of the quarry, but fortunately no lives were lost as the fall occurred in the night. There is nothing of particular interest now until we come to the beautiful section of rock at Coniston Railway Station before noticed.

Proceeding through the village of Coniston and bordering the Park at Coniston Waterhead, we come to the excellent slate and flag quarries belonging to J. G. Marshall, Esq., always in active work, and producing very good material; the flags are sawn and dressed by machinery in a superior manner. Very few Organic Remains have ever been found here. The Coniston Flag at this point occupies not only the whole space from the quarries to the lake, but extends at least a mile down the east side of it, giving a breadth of upwards of two miles, increasing as we proceed eastward towards Windermere Lake, and as this formation dips here at a high angle, it must be of immense thickness. After leaving Coniston

Waterhead, we have a steep ascent to Hawkshead Hill, near which place three porphyry dykes crop out in the road; one of them, according to Professor Sedgwick, may be traced for a great distance both on the north-east and south-west of the public road leading from Coniston to Hawkshead.

About a mile east of Hawkshead Hill, are the slate and flag quarries of Cold Well, noted for producing some splendid specimens of Organic Remains, particularly the *Odontochile Obtusa caudata*. We have obtained some noble specimens of this strange crustacian at Cold Well. The Organic Remains are not interspersed through the whole mass of rock, but only in one single "fossil band" which dips with the plane of deposit at an angle of about 35°, and may be clearly traced on the face of the rock from the surface of the ground obliquely to the bottom of the quarry, so that none can be had except the workmen are engaged on this part, but when a blast is made in the "fossil band" it would delight any geologist to see large slabs brought down, each one having a face almost covered with the heads and carrapaces of these splendid Trilobites. Unfortunately they are never in contact, and we can rarely determine which head and body belong to the same individual fossil. Besides the *Obtusa caudata*, there are other species in a high state of preservation, a list of which will be given in the appendix.

Cold Well quarries are without exception the best localities for Organic Remains in the whole range of the Coniston Flag formation. Two miles north of Cold Well, are the celebrated flag quarries of Brathay, the last open quarries in the Coniston Flag, and near to its north-east boundary. At a short distance beyond Brathay we trace this beautiful rock formation into Windermere Lake, consequently we lose it entirely in Furness. The Coniston Flag is an excellent material for all building purposes, being sound and durable; but in no part of its range can it compete with the instructive quarries at Brathay.

A curious example of jointing is seen in these quarries, viz., that the flag metal is perfectly free from seams or joints of any kind for a considerable length in one direction, and at right angles to this there are seams as straight as a line can be drawn, which when opened show a face covered with beautiful crystals of iron pyrites, giving it the appearance of being plated with gold. These blocks will split up into stone planks ten or twelve feet long, two or three inches in thickness, and from twelve to eighteen inches in breadth, with clean and beautiful self edges plated with pyrites. Many of these are sent to distant places and used for garden edgings and other ornamental purposes. These quarries have yielded very few Organic Remains. The *graptolites priodon* are sometimes found in the nodular concretions, with which these interesting quarries abound. We have now ended our second traverse across Furness, and must return to the estuary of Duddon, near Foxfield Railway Station, to take the Coniston grit, as it is the next formation in the series of deposit.

CONISTON GRIT.

Our third traverse across Furness will be parallel to the first and second, viz., from south-west to north-east, on the Coniston Grit, the next formation in the series of stratification above the Coniston flag, and the base of Professor Sedgwick's Silurian system. The Coniston Grit*

---

* A thick group composed of hard siliceous sandstone, in some places of very coarse texture, and passing into a conglomorate form. It is very steril of fossils, and the few which have been found give no decisive evidence as to its epoch. It much resembles the steril portions of the Mary Hill sandstone, and it appears to have the same place in the general series.—*Sedgwick.*

is a hard siliceous sandstone, generally flaggy and broken, and very little used for building purposes, except for fence walls. This formation first shows on the surface, near Foxfield Station, on the Furness and Whitehaven Railway, its south-west limit; and although its range is parallel with the Coniston limestone and the Coniston flag, it does not, like them, extend across the estuary of Duddon into Cumberland, but is covered up here and appears no further to the south-west.

A short distance from Foxfield Station there is a quarry in the Coniston Grit showing a bold escarpment of rock, the face of which presents a curious waved and knotted appearance, and the knots being pear-shaped and easily detached from the rock, have been mistaken for fossils, but they have not the slightest trace of Organic structure. The breadth which this formation occupies above ground at this end of its range is about a mile and a half, but its junction with the Lower Ireleth slate on the south-east, cannot be determined on account of the extensive Angerton mosses in that direction, and as the whole of Angerton Farm and Waitham Hill are Coniston Grit, it is evident that its junction with the former stratum occurs somewhere between Waitham Hill and the east side of the mosses near Head Crag. About a mile north from this quarry on the line of strike, viz., at Walls End Farm, there is a small quarry in the Coniston Grit, the produce of which is used for fence walls, being very fragmentary and unsuitable for any other purpose. Two or three small fragments of Organic Remains have been obtained at this place, but the species could not be determined, and we have picked up from the private carriage-road leading to Eccleriggs, the seat of R. Assheton Cross, Esq., M.P., three or four different species of Organic Remains, which will be noticed hereafter. The interior of these extensive mosses well deserves the attention of the botanist, but it is not much visited on account of the difficulty of transit, as there are large plots of soft and even dangerous ground, the substratum being a semi-liquid mass below the tough sward on the surface. We noticed two or three curious species of aquatic plants and several strange species of moss, but the *Osmunda regalis* was most abundant. In some allotments where the peat had been taken away, and the land lowered in consequence, the place was almost literally covered with this beautiful fern; a thousand specimens might have been collected in the space of half an acre. Some were very luxuriant, others dwarfed to five or six inches in height, yet a few even of the smallest specimens carried the distinctive flowering seed stem of the Royal Fern. About half a mile north-east from Walls End, in the moss land belonging to the same farm, a porphyry dyke bursts through the moss and extends on the surface about 150 yards. A portion of it so is highly decomposed that it may be dug out with a spade as easily as if it were sand. The colour inclines to yellow, from having innumerable particles of bright yellow mica interspersed through the whole mass, which circumstance induced a party to make an excavation in it, expecting to find copper ore, or perhaps something more valuable.

This singular outburst of igneous rock in the centre of these extensive mosses has hitherto escaped the notice of geologists, and as there are no marked geological features which can be seen from a distance, it offers no attraction to strangers, though we think it is well worth a visit. From Walls Moss it ranges north-east through Woodland and Torver, to the western shore of Coniston Lake and a great portion of the rough rocky hills on the south side of Torver, between Sunny-bank and Water-yeat, belong to this formation.

These hills may be considered the north end of the Kirkby Moor range, which commences near Ireleth and taking a northerly direction by Gillhead, Gawthwaite, Burney, Todhillbank Fell, Beacon Tarn, ends a short distance to the south of Torver Chapel. From Todhillbank Farm to the north end of the ridge, the district is exceedingly rocky and steril, and except the porphyry dykes and the patches of limestone near the latter farm, already described, there is nothing of geological interest.

After crossing Coniston Lake there occurs a fault similar to that of the Coniston limestone, near Mr. Marshall's, at Coniston Waterhead, which throws it about a mile out of its ordinary course, and it is a remarkable fact that both this and all the other faults affected by this dislocation, change their bearing, but have a tendency to return again to their original north-east direction, after which their course continues as before. From the eastern shore of Coniston Lake the Coniston Grit ranges by Grizedale Head, the north end of Esthwaite Lake and Latterbarrow to Windermere, from the eastern shore of the last by Elleray, thence over Applethwaite Common by the south end of Kentmere Tarn, and the next ridge of hills passing to the north of Longsleddale Chapel. Before we arrive at this point the grit formation is changed in colour, and the pale red rock seen near Foxfield Railway Station has now become a grey gnarled stone, hard and more intractable than when first seen at its south-west end at Foxfield, and resumes its course in a north-east direction by Shap Wells to the waterfall of Cautly-spout, beyond which it becomes degenerate and fragmentary.

Organic Remains are very rare in all parts of this formation, yet Professor Sedgwick gives the following list as having been obtained in this group of siliceous rock, viz.: *Graptolites ludensis*, *Cordiola interrupta*, *Orthoceratites Ibex*, *O. subundulatus*, and fragments of *Trilobites, &c.* All the known species are Upper Silurian, and to this list we have added two or three others which will be noticed hereafter.

In Professor Sedgwick's first supplemental letter to Wordsworth (Page 221 of Wordsworth's Guide to the Lakes) he gives an ascending section through the fossiliferous slates of Westmorland, Furness, &c. In this section he divides the Ireleth group into four members, viz., 4a, Lower Ireleth slate; 4b, Ireleth limestone; 4c, Upper Ireleth slate; 4d, coarse slate and gritty, which in his great work on the lake district he also calls Lower Ludlow rock. In 4c of this group the Professor found the following species of fossils, viz.:—*Graptolites ludensis a cyathyllum*, *Favocites alveolarus*, and two or three *Orthoceratites*, and in 4d occur, though very rarely, corals, encrinite stems, and *Cardiola interrupta*; all the known species made out are Upper Silurian. It is evident, for reasons already given, that his classification of the Ireleth group of slate rocks was not considered even by himself as a finished work, but only provisional, as at the time of the Professor's survey of this part of Furness very few fossils had been found in the slate rocks of Kirkby Ireleth. Since that period we have obtained a very good list from the Kirkby Moor range, all of which are characteristic fossils of the Coniston flag, and when the specimens obtained from Ash-ghyll are placed side by side with those from Kirkby Moor in the same cabinet, they cannot be distinguished from each other. Moreover, these Organic Remains have not been found at one particular quarry or place alone, but at all the quarries on the Kirkby Moor range, from the trial quarries at the south end of the moor to Stone-dykes in Lowick on the north, viz., Harlock, Knotthollow, Iron Yeats, Groffacragg, Scars, Parkgate, Gawthwaite, and Stone-dykes.

LOWER IRELETH SLATE.

The next formation above the Coniston grit in the general series of stratification is the Lower Ireleth Slate. Crossing the extensive mosses from Foxfield in a south-east direction to Kirkby Ireleth, we also cross the junction of the Coniston grit and the Lower Ireleth Slate, which is somewhere under Angerton Moss; but we have no *data* for asserting even approximately where it occurs, although the junction certainly is somewhere under the mosses between Waitham Hill and Head Crag, as we trace the Coniston grit under them on the north-west side, and the Lower Ireleth Slate on the south-east. The latter may be said to commence at its south-west end on the shore of the estuary of the Duddon, near to the village of Soutergate, in Kirkby, taking a direction about north-north-east, by Sandside, at which place there is an open quarry unused for several years. At this point it is about half a mile in breadth, or generally it occupies all the low land, and also all that portion of the hill-side nearly up to the great quarries of the Upper Ireleth Slate, which extend along the western slope of Kirkby Moor for nearly a mile. From this point northwards, by Grizebeck and Kirkby Park, the breadth increases rapidly so that near Causeway End it extends the whole distance from Angerton Moss eastwards to beyond Burney Farm, thence northwards over Heathwaite Fell to Todhillbank Farm, and occupies the whole of the rocky district to its junction with the Coniston grit in Woodland. Near Todhillbank, two or three porphyry dykes crop out, one of which may be traced to the foot of Coniston Lake, and from there, after crossing the water, two or three miles further in a north-east direction. Professor Sedgwick supposes all these dykes may proceed from one centre of disturbance. About half a mile north-east from Todhillbank Farm, several bluffs of limestone rock crop out, which the Professor supposes to be a continuation of the Ireleth limestone, and from them he obtained encrinites with a four-sided column. We also found several other fragments of Organic Remains so mutilated that the species could not be decided.

We have spent two or three days at this place, without succeeding in finding the four-sided encrinite, but after long and patient working we secured three or four different species of Organic Remains so nearly perfect that the species could be determined with certainty. These are all characteristic Coniston limestone fossils, and nearly all the fragments found were of the same type, therefore it is probable from these facts the Professor will pronounce this outburst of limestone rock to be an outlier of the Coniston limestone, similar to that at High Haume.

From this point eastwards, the Upper and Lower Ireleth slate seem imperceptibly to blend and lose their perfect "slaty cleavage," so as to be undistinguishable from the Lower Ludlow rock of Hoad and Outrake, which is apparently their general character through the whole of the rocky district, by Coniston Water Foot, Rusland, &c., to the western shore of Windermere Lake, extending on the shore of the lake southwards, from the Ferry to Newby Bridge, and through the greater part of its range, occupying a breadth of several miles.

IRELETH LIMESTONE.

This unimportant member of the rock formations of Furness is next in the series, in the ascending order, and is first seen in some trial quarries, in several fields on both sides of the road leading from Ireleth to Kirkby, none of which are now in work. At High Mere Beck, a few years since, there was a quarry in this limestone from which, after diligent search, we obtained a few imperfect specimens of Organic Remains of two or three different species, all of them of the type of the Coniston fossils, but nearly of the colour of the mountain limestone. This quarry is now filled up, and a limekiln which was attached is cleared away entirely, so there is not much evidence on the surface that the Ireleth Limestone exists, although it was more largely developed at High Mere Beck, than any other place.

The range of the Ireleth Limestone is somewhat different from that of the lower stratifications of Furness, its true magnetic bearing being north-north-east, *i.e.*, from the north side of the village of Ireleth, it skirts the foot of the hill range of Kirkby Moor, by Moorside, Tippin's Bridge, to High Mere Beck, thence gradually rising the breast of the Kirkby Moor range to pass on the east side of Brig House, Bayliff Ground, Gargrave, and Bank House, where it is seen again on the surface in Gill Wood, beyond which it shows again in a small trial quarry in one of the fields belonging to the Low Hall Estate, and again in the beck-course above Kirkby Mill. This appears to be its extreme northern limit in Kirkby, and from the ascertained thickness in several parts of its range, at all which places it gradually thins as we proceed northwards, it will end with a fine point near Banks Gill Beck, proving that this mysterious deposit of Ireleth Limestone has the appearance of a wedge driven in between the Upper and Lower Ireleth slate.

We say mysterious, because from its colour and mechanical texture, we should pronounce it to be mountain limestone, yet we have the strongest fossil evidence that it belongs to an older, and consequently a deeper zone in the stratification in the earth, showing in the present instance mechanical texture and colour, as antagonistic to fossil evidence, which is a circumstance of rare occurrence. Our own opinion is that the whole of it is Coniston limestone, and this is strengthened by evidence entirely distinct from any afforded in this locality. Professor Sedgwick, who is acknowledged to be a very high geological authority, says, in his " Geological Letters to Wordsworth," that " two or three patches of Ireleth Limestone crop out on Todhillbank Fell, which contain Organic Remains, but all of them so fragmentary that the species could not be determined." Having, as before stated, devoted a considerable portion of time and much patience in thoroughly examining all the places indicated by the Professor, we succeeded in finding three or four species of Organic Remains so nearly perfect, that the species could be determined with certainty, and all proving to be characteristic Coniston limestone fossils, we have no doubt whatever that the patches on Todhillbank Fell, as well as the whole band described above, are the Coniston limestone.

LOWER LUDLOW ROCK.

The preceeding chapters have all been devoted to the examination and description of stratifications, for the most part with clearly defined boundaries. We will now notice the last division of the clay-slate formation in Furness, —the southern, or that which forms a junction with the great deposit of mountain limestone crossing the peninsula of Furness, from west to east, *i. e.*, from the east shore of the estuary of Duddon, near Dunnerholme, to the west shore of Morecambe Bay, at Plumpton. The true interpretation of this division of the clay-slate may be considered one of the most difficult chapters in the whole science of Geology, and we have some diffidence in attempting to deal with it in any way, although we feel we ought not to pass it over without a few observations, however imperfect they may be.

One source of difficulty arises from its fragmentary character, as no continuous natural section can be constructed a mile in any part of its range, and it is not only vandyked, or toothed, into the mountain limestone in a strange manner on its southern boundary, but is also interrupted and protruded through on the northern and middle portion of it, by igneous rocks of various kinds, viz., porphyry, amygdaloid, green slate, &c. On the High Haume Estate there are outbursts also, of the lowest stratified rocks of Furness, the Coniston limestone, Coniston flag, &c., forced through the superincumbent members of the Lower and Upper Ireleth slate, and brought to the surface to be interspersed among masses of all the igneous rocks enumerated above, and at one point, near High Haume, the Coniston limestone, and the Mountain limestone, are almost in actual contact. This protruded mass of Coniston limestone is highly fossiliferous, and contains many species of Organic Remains identical with those of Coniston Waterhead, two or three species appearing to be peculiar to the place.

Another source of difficulty arises from the fact of this part of Furness having formerly been considered almost entirely unfossiliferous, even Professor Sedgwick, the great geologist, whose untiring labours have done more to interpret this complicated district than all other geologists together, laments that "the physical history of this district cannot be sufficiently made out from its great paucity of fossils, nay," says he, "it is almost without fossils."—(See

Sedgwick's Geological Letters to Wordsworth). The different rock formations of Furness were therefore classed, and received such names as were due to the mechanical texture and the same chemical composition with known rock formation in other districts. Thus, the great quarries of Kirkby were, by Professor Sedgwick, called "Upper Ireleth Slate," a local and very appropriate name, and from about the summit ridge of Kirkby Moor eastward towards Ulverston and Lowick, he called "Lower Ludlow Rock," and at Rosshead, near Ulverston, a small patch of Mud-stone occurs, described by Mr. Salter, as "Wenlock Shale." Abundant Organic Remains have recently been discovered by local geologists in this division of the clay-slate formation, not only at Rosshead, but in several other and unexpected localities, therefore, a new classification of the rock formations of the district becomes necessary, founded on the combined evidence of mechanical structure and Organic Remains, and it is highly probable that the former opinions of geologists will be somewhat modified by new facts which will be brought before them. There is one fact of great importance which requires especial notice, as it interferes materially with Professor Sedgwick's classification of the rock formations of Furness. In his letters to Wordsworth he says, "The Coniston grit may be considered as the true base of the Silurian system, as above it (the Coniston grit) a new type of life prevails, and the stratifications above the grit scarcely contain a single species of Organic Remains, which are found in the stratification below it, and only four or five per cent. are common to both." Now, the Ireleth slates are above the Coniston grit, and the Coniston flag below it, and although we have found a great many species of Organic Remains, in several different parts of the Kirkby Moor range, all of them were characteristic fossils of the Coniston flag, and identical with those we have obtained from Lords Quarry, on Broughton Common, from Ash-ghyll, in Torver, and from Cold Well Quarry, near Brathay, viz. : — *Cordiola interrupta*, *Obtusa caudata*, &c.

In fossil evidence therefore, we have a perfect agreement between the Ireleth Slate of Kirkby Moor and the Coniston Flag of Broughton Common and Ash-ghyll, and with reference to mechanical structure and colour there is a greater difference between the slate from one quarry on Kirkby Moor, and that from another on the same moor, than there is between the slate of Kirkby Moor and the slate of Ash-ghyll or Broughton Common ; thus, in fossil evidence, mechanical texture, and colour, they exactly coincide, and prove that the Coniston flag and the Ireleth slates are two different members of the same formation with the Coniston grit intercalated between them.

The junction between the clay-slate and the Mountain limestone from a point near Soutergate, in Kirkby, tends southwards between the Furness Railway and the east shore of the estuary of Duddon, and maintains a straight course to the west end of Askham Wood. This line we may assume theoretically, to be the boundary between those two formations, and although the bold headland of Dunner-holme jutting into the estuary of Duddon, is the northernmost Mountain limestone rock seen in Furness, it is highly probable that it not only extends further to the north, but also to the south-west under the estuary to Hodbarrow, and also to Limestone Hall, near Silecroft, in Cumberland. From Askham Wood southwards, the boundary between the Upper Ireleth slate and the Mountain limestone ranges with the Furness Railway down the valley, and near the western base of Greenscow Wood, it abuts on the igneous rocks of High Haume, skirting round the base of the hill by Mouzell to Orgrave, where we come upon a small patch of old red sandstone in the brook near Orgrave Mill, this sandstone is tinted of a greenish drab colour, and accompanied by the usual old red conglomerate. The northern boundary of the limestone from this point ranges northwards up the valley by Holmes Green to Scale Bank, where it again nearly comes in contact with the Coniston limestone of High Haume. From Scale Bank, the junction between the Mountain limestone and the clay-slate, is nearly a straight line to Marton, and passes through the centre of the village, some of the houses standing on the clay-slate, others on the limestone.

At Old Hills, near Marton, there is a small quarry in the mountain limestone, where we have found the *Mitchliena grandis*, and several other species of Organic Remains characteristic of this formation. It has also produced several species of splendid vegetable fossils, peculiar to the coal formation, viz., *Sigilaria, Stigmaria, Lepidodendron, Calamites, &c.*, many of the specimens being eight or nine inches in diameter, and three feet in length. This is the only Mountain limestone quarry in Furness, with the exception of traces at Tarn Close, where vegetable fossils have been found. From Marton, the boundary between the clay-slate and the Mountain limestone ranges north-east to Snipe Gill, or Waste Cottages, thence along the brow of the hill southwards to Carkettle, the site of the iron ore works of J. Rawlinson, Esq., where in some of the underground drifts or galleries the junction between the Mountain limestone and the clay-slate is very palpable. Valuable veins of ore occur in the limestone where it abuts upon a cheek of clay-slate, but in no part of its range does the ore penetrate into the slate rock as it does into limestone, filling cracks and "ginnels" in the stone like molten metal filling a mould, thus giving one class of evidence respecting the age of our hæmatite ore, that it has been deposited more recently than either of the rock formations with which it is in contact.

It is by no means singular to find iron ore in a situation like this, neither is it the only instance occurring in our district, but is in accordance with a well established theory, that metallic ores may reasonably be expected at the junction of any two rock formations of different mechanical structure and different chemical composition. In tracing the extent of the limestone, from Carkettle south-west to Lindal Moor and Whinfield, we almost follow the boundary between the Parishes of Pennington and Dalton, which also divides the Manor of Pennington, (the property of Lord Muncaster), from the Manor of Plain Furness, (the property of the Duke of Buccleuch).

We are now on the site of one of the greatest deposits of hæmatite ore in Britain, of the richest quality, and although it has been wrought here for hundreds of years, yet, within our own remembrance, only half a dozen old men found employment in scratching up a little of the best of it, leaving for the present and future generations to raise and dispose of millions of tons of the valuable mineral. These works have been carried on for many years by Messrs. Harrison, Ainslie, & Co., who now employ several hundred miners constantly. The pits of Lindal Moor produce many beautiful specimens of rich iron ore of the "Kidney" or Reniform structure, some of them so highly ornamental, that they serve to adorn the mantel shelves of most of the cottages in the neighbourhood, but some of the good wives, not content to display them in their natural beauty, give to their specimens a coating of black lead, which, however, does not improve their appearance. Good specimens of crystallized carbonate of lime, may also be obtained here, many of them semi-transparent, and the miners collect them for ornamental purposes; but the most singular form of structure assumed by the hæmatite ore, consisted of naturally formed pencils or marlin-spikes, which were found about three years ago, in excavating a "drift" or gallery in a vein of very pure ore, many of the pencils being seven or eight inches in length, and from a quarter of an inch to an inch in diameter, perfectly smooth and polished on the outside, and of a fibrous texture within, most of the fibres as small as human hair. This deposit occurred in a pit called B.29, the quantity found being only small, not more than one or two tons. When one of the blocks of ore containing pencils is viewed in section, rings are clearly defined in the mass, representing the bases or thicker ends of the pencils, and a smart blow on the end of the block will disengage some of the spikes, which can then be drawn out like daggers from their sheaths. B.29 pit is still in active work, but no trace of this singular structure of ore has been found, either before or since that time. The ore of Lindal Moor has afforded some facts having a direct bearing on that mysterious question, the formation of our hæmatite ore.

We have now lying before us a fossil of perfect iron ore, a portion of the column of a species of encrinite, an inch and a half in length and five-eighths in diameter, possessing the same specific gravity as the hæmatite of the richest quality. When it was found, it enjoyed the distinction of being the only iron ore fossil ever brought to light in our mines. Subsequently, however, we found two or three others which seemed to be in a transition state, being neither perfect iron ore nor perfect limestone. These are all marine fossils characteristic of the Mountain limestone. W. Ainslie, Esq., (of the firm of Harrison, Ainslie, & Co.) has also found a complete iron ore fossil, *Producta marina*, but we are not aware that any others have ever been discovered. The circumstance that Organic Remains have turned up in hæmatite ore is very important, as being a demonstration that some portion at least, if not the whole of our hæmatite ore, is of aqueous origin. After leaving Lindal Moor, in tracing the northern boundary of the Mountain limestone, we proceed eastward through Whinfield, pass Gill-brow on the north side of the hill, also the north side of the new iron ore works of J. Rawlinson, Esq., near Cross-a-moor, through the village of Swarthmoor, thence towards Swarthmoor Hall, from which place we cannot trace it with certainty eastward, until we come to Swarthdale House, the seat of George Sunderland, Esq., where it may be seen *in situ* in the bottom of the beck. Eastward of this, there is some uncertainty in its range as there is no rock on the surface for a considerable distance, but we shall approximate to it by following the beck-course through Dragley Beck, thence by the Low Mill to the Outcast, where we leave the streamlet altogether and take a straight line north-east across the Ulverston Canal, near the gunpowder slip, to the north end of Hag Spring Wood, in Plumpton, where it may be traced with certainty for a considerable distance, as it ranges between Hag Spring and Flushes Woods, and in the adjoining field, clay-slate and Mountain limestone rock both crop out on the surface within one hundred yards of each other, the Ulverston and Lancaster Railway passing between them. From thence it ranges eastwards to the shore of the Leven Estuary, near to Plumpton Hall, where it is lost under the sands of the bay, but appears again on the opposite side of Ulverston Sands, and skirts the shore with a narrow belt for about three miles, from Holker Park to the north end of Roudsea Wood.

It will be seen from the above traverse that the line of junction between the Mountain limestone and the clay-slate is not a perfect one, but the two formations are strangely indented into each other, so as to make it very difficult, and in some parts of the course impossible, to clearly define the boundary between these two highly important depositions. This is especially the case with

LOWER LUDLOW ROCK.

reference to a deposit of Mountain limestone at Rosshead, which also extends into the neighbouring estate of Tarn Close, where Messrs. Rawllnson and Briggs had iron ore works. These works are now given up, but the site is still occupied as a limestone quarry, which produces stone of a finer texture than that either from Birkrigg or Baycliff, or any other part of the great limestone deposit of Furness. As no rock has been found between Tarn Close and the Swarthmoor Hall Estate, it cannot yet be determined whether this deposit is an outlier, or island, of limestone, detached from the great body of Furness limestone, or, if it is continuous from Rosshead to Swarthmoor new village on the south-west, also from Tarn Close, by Fair View, the seat of Mrs. Kennedy, to Swarthdale House on the south-east, thus forming a great mass or horn of limestone, protruded between the clay-slate of Fell Side and Channel House, in Pennington, on one side, and the clay-slate of Gill Banks and Lightburne Park, on the other. Now, although we have no direct proof either way, yet there are circumstances which make it highly probable that the limestone of Tarn Close and Rosshead, forms a junction with, and is part of the great Mountain limestone deposit of Furness.

One reason for the above opinion is the fact that in the main road at Rosshead, there crops out a soft mud-stone, which Mr. Salter calls "Wenlock Shale," but which, in truth, is an altered condition of clay-slate.* It decomposes and wastes by atmospheric influence, with greater rapidity than the hard clay-slate of Gill Banks and Lightburne Park, and the equally hard clay-slate of Fell Side and Channel House, so that when the ancient Silurian sea washed the base of our clay-slate hills, as Hoad, Outrake,

---

* The mud-stone of Rosshead abounds with beautiful Silurian fossils of different species, particularly two or three species of *Cardiola interrupta*, *Pterinea Lingula*, &c., &c., which will be noticed hereafter.

Flan Hill, Gamswell, Whinfield Point, and Kirkby Moor, —all prominent objects for the south and south-western storms to break upon, before the deposition of the Mountain limestone,—the mud-stone of Tarn Close and Rosshead would be abraded and wasted, so as to form between Ulverston and Whinfield Point, a deep bay, which was subsequently filled with liquid carbonate of lime.

Again, between High Haume and Whinfield Point, there would be another bay, extending to the village of Marton, where the soft Powka-beck stone has been decomposed in a similar manner, and the whole bay filled in with carbonate of lime as far as Marton and Snipe Gill. The above statement does not admit of any doubt, as we have direct proof that such was the fact, yet the great mystery still remains how our hæmatite ore became mixed and blended with Mountain limestone, introduced into this Silurian bay, for it is the site of the greatest deposit of hæmatite ore at present known in Britain, comprising the mines of Cross Gates, Whitriggs, Old Hills, Carkettle, Lindal Moor, Ure Pits, and Lindal Cote. A still greater marvel is propounded in the question of the formation of iron ore itself, with which none of our great geologists has hitherto attempted to grapple. We are happy to say, however, that this is likely soon to be remedied, as at least one master-mind will be brought to bear on the subject, for, in a conversation we had in July, 1865, with Professor Phillips, of Oxford, who was then at Wastdale Head, collecting rock specimens, we arranged to bring before him on his return, such facts as seem to bear on the ore formation, and especially direct his attention to the *debris* from one of the northern pits of Lindal Moor, which offers some new data for reasoning on this very obscure subject.

We have now lying before us a specimen of this *debris* from the north-east side of the Lindal Moor deposit, which shows true alternating layers, each about two inches in thickness, of perfect iron ore, and perfect Mountain limestone white and untinged with iron, all the layers forming junctions so complete as to give the idea that the limestone and the hæmatite were both in the condition of liquids at the same time. Another example shows streams of iron ore running into a mass of limestone where there is not the least appearance of a crack or vein, which the hæmatite subsequently filled up, thus affording some slight evidence against the supposition that the deposition and hardening of the Mountain limestone occurred before that of the hæmatite ore commenced.

It has hitherto been the accepted theory that our hæmatite ore is later or newer than the Mountain limestone, but the great deposit of iron ore at Lindal Moor, affords some testimony that the ore was deposited first, and the limestone afterwards, because there, as well as at the works of J. Rawlinson, Esq., at Carkettle, the ore lies between the Silurian clay-slate and the Mountain limestone, showing that the latter is not so old as the hæmatite ore. At the same time there is contradictory evidence of the case being precisely opposite, and, moreover, that they were both in a liquid state at the same time. We may therefore well say that the formation of our hæmatite ore is veiled in obscurity, and we have not attempted to unravel this knotty point, our remarks hitherto having reference only to the relative age of the two deposits, the graver difficulty we leave to abler hands.

We have also a specimen from the iron ore works of Alexander Brogden, Esq., M.P., at Stainton, which shows a perfect junction between the Mountain limestone and hæmatite ore, the latter assuming the columnar form, somewhat resembling the basaltic columns of Fingal's Cave, in the Isle of Staffa. The columns, which are rough and harsh to the touch, and look as if they had been subjected to heat, rise from a flat surface of white

Mountain limestone unstained with iron, and the union is so exact that the two formations cannot be separated without breaking one or both to pieces.

The specimens showing alternate layers of iron ore and Mountain limestone, are not choice or select, but were taken without much care from the mass of *debris*, consisting of several hundred tons, now lying near the mouth of the pit at Lindal Moor, from which it was excavated, and this rubbish, utterly worthless for economic purposes, will afford a valuable lesson for the geological student. The facts detailed are in themselves curious and interesting, and may be taken as elements for reasoning when dealing with this important but complex subject.

We have also before us a class of specimens from Mr. Brogden's iron ore works, at High Kinmont, near Bootle, in Cumberland, which teach a very different lesson. These works are in granite, decomposed to such an extent in some places, that it may be dug out with a spade, and used as sand, in other parts of the works there are all the different gradations of decomposition, from loose sand to perfect hard and sound granite rock. One portion of the rock is quite free from iron, the next is slightly reddened, another is more deeply imbued, a fourth has a film or coating of iron ore on the outside, with particles of iron and disintegrated granite interspersed through the whole mass, and the last is a specimen of perfect iron ore. These are suggestive, but we offer no opinion, merely mentioning the facts for wiser heads to decide upon.

We have already stated that the Silurian sea at a very remote period of the earth's history, broke against the base of the clay-slate hills of Furness. This may appear a startling assertion to those who have not read or studied geology, nevertheless the science offers positive proof that it is true, for as sure as the sea now washes the shores of Low Furness, so surely did it break against the base of

Hoad, Outrake, Flan Hill, Gamswell, Whinfield Point, and Kirkby Moor, at which period *Low Furness* had no existence on earth, but was afterwards deposited by slow degrees in a state of semi-liquid carbonate of lime, beneath the waters of the sea. After a lapse of countless ages, Birkrigg, Baycliff Haggs, and Stone Close at Stainton, appeared above the waters as small islands of limestone, and last of all the newest rock formation in Furness, the Upper Permian sandstone of Hawcote and Furness Abbey. It is a curious matter to speculate on the age of a rock, as geology does not give any exact chronology, yet it shows us certain epochs, or stages, in the duration of time, by which we can compare the age of one formation with another in distant parts of the earth, for the order of the deposition of all the different stratifications follows a certain well established law, but if we attempt to estimate even the newest of them with the oldest remains of the works of man on earth, the comparison is ridiculous, being only as a second of time to a thousand years. For instance, we naturally associate the idea of great antiquity, with the "Royal Abbey of Saint Mary in Furness," but it may somewhat detract from our veneration for its age, when we consider that, according to the law of superposition indicated above, the stone with which the abbey is constructed, was not in existence for thousands of generations after the deposition of the Silurian rocks of Hoad, and other clay-slate hills in the parishes of Ulverston, Pennington, and Kirkby Ireleth. Our antiquarian friends, however, need not be uneasy on that account, for the same law tells us that the stone used in building the Pyramids of Egypt (the Numulite limestone) was not in being for countless generations after the deposition and consolidation of our comparatively new rock, the Permian sandstone of Hawcote and Furness Abbey. Again, even this Egyptian rock was formed thousands of ages before the Almighty Creator and Ruler of the Universe, in His own good time, and for His own wise purpose, placed man upon the earth.

Furness offers many geological facts of high and enduring interest, especially that portion of it now under our consideration, for here we have inexhaustible quarries of slate and lime, of the very best kind, and deposits of valuable hæmatite iron ore, unquestionably in the greatest quantity and of the purest quality in Great Britain, yet it is very probable one-tenth part of it is not yet discovered, for our enterprising mining companies are finding new "lodes" or veins of ore in places where the old miners would have thought it madness to seek. Some of those places gave no surface indications whatever that iron ore in abundance was lying beneath, for even the far-famed Lindal Moor lode has not in the least degree tinged or reddened the soil of the superincumbent adjoining land. From which we may reasonably infer there may be discovered hereafter many other veins, sops, or basins, filled with ore, when the district of Low Furness is thoroughly proved by "borings" or other suitable means.

This small insulated district certainly seems to be peculiarly favoured by Providence, for we have not only a fruitful land, crystal rivers, pure air, and beautiful scenery, but an unlimited store of riches under our feet, and a rural population even in the mining district of Low Furness, which will bear a favourable comparison with any part of England, while the dalesmen and shepherds of the mountainous parts of High Furness, would be difficult to equal for manliness, truth, and honesty, anywhere in Britain.

MOUNTAIN LIMESTONE.

The northern boundary of the Mountain Limestone, which crosses the promontory of Furness, detailed in the last chapter, (p. 76), may be accepted as a close approximation to its true range, for it is the result of several careful surveys undertaken expressly for the purpose of determining its junction with the clay-slate underlying it on the north side through the whole course of its range, and with the exception of that portion of limestone rock, at Tarn Close and Rosshead as before stated, there is nothing doubtful respecting it.

We will now attempt to follow the southern boundary of the Mountain Limestone also ranging from the east shore of the estuary of Duddon to the west shore of Morecambe Bay entirely across the promontory of Furness, underlying and forming a junction with the Upper Permian sandstone of Hawcote and Furness Abbey, which, however, cannot be traced through the whole line of the Mountain Limestone eastward, although they range together as far as the former can be made out to the south-east. It is highly probable, nay, almost certain, that the great Mountain Limestone deposit of Furness, does not actually commence in Furness, but about three miles to the westward at or near Limestone Hall, in Cumberland, which is about half way between Sylecroft and Kirksanton. From this point westward it certainly does not extend to Sylecroft, but is cut off by the Skiddaw slate, of Black Combe, and the great fault which ranges up the valley of Whicham, therefore Limestone Hall determines the limit of the Mountain Limestone both in a northerly and westerly direction,. as it forms a junction with the green slate and porphyry of the Millom Park range, within two hundred yards of the Limestone Hall quarry. From the latter we have obtained several species of Organic Remains, all characteristic fossils of the Mountain limestone, comprising several species of the genus *orthis, producta, spirifer, &c.*

From Limestone Hall, the northern boundary of the limestone deposit courses eastward by Kirksanton and Hestham, thus far abutting on the south-west end of the green slate and porphyry of the Millom Park range, and at this point we nearly come in contact with the Coniston limestone of Beck Farm, in Millom, while on the south side of Holborn Hill, it forms a junction with the south-west end of the Coniston flag, then probably takes a north-east direction across the estuary of Duddon, to join the Furness deposit of limestone on the north side of Dunnerholme. Again, from Limestone Hall, the limestone deposit widens out to the south by Standing Stones* towards Haverigg, thence skirting the sea-shore by the New Hall Estate, to Hodbarrow Point, and crossing the estuary of

---

* Standing Stones, or "The Giant's Grave," is a monument of undoubted antiquity, which has apparently escaped the notice of our antiquarian friends. It consists of two immense stones fifteen feet apart, each several tons in weight, erected in one of the fields of this estate, the larger being ten feet, and the smaller eight feet, above the ground. They are entirely undressed, similar to the blocks forming the "Druid's Temple," at Swinside, only much larger, and are of the same geological formation, the green slate and porphyry. It is singular that all those monuments in the Lake District, known as "Giant's Graves," are fifteen feet in length, one on Heathwaite Fell, between Burney and Woodland, was opened a few years since, by the late Rev. Francis Evans, of Ulverston, and several fragments of charred human bones and a stone ring were found, showing that a human subject had been buried there; but the most celebrated is that in Penrith Church Yard, which is also fifteen feet in length, and besides having a high stone erected at each end, has edging stones placed all round it. This is said to be the grave of the legendary giant, the monstrous "Owen Casarus," who lived in a cave on the banks of the river Eamont, and committed great depredations amongst cattle and sheep, for many miles round, and the old women in this part of Cumberland, tell how he sometimes came home to his retreat with a dozen sheep, their heads under his belt, strung dangling round him. This interesting relic has doubtless given name to the estate, and it was very likely in existence when all the district was a wild and desolate waste, many hundreds or even thousands of years before the land around became enclosed for agricultural purposes.

Duddon, at a considerable distance to the south of the large iron ore works of the Hodbarrow Mining Co., to Roanhead, on the west shore of Low Furness.

It will thus be seen that the Mountain Limestone formation makes a fringe or belt between the Furness and Whitehaven Railway and the sea-shore, all the way from Limestone Hall to Hodbarrow Point, abutting on the south-east ends of three important geological formations, Green Slate and Porphyry, Coniston limestone, and Coniston flag, all of which extend in a north-east direction without any interruption whatever for upwards of fifty miles. This border of limestone in no part of its range exceeds one mile in breadth, and may be considered as a sort of tail appended to the great Mountain Limestone deposit of Furness, which after passing Roanhead, begins to widen out suddenly, its southern limit extending almost to the farm of Sinkfall, thence eastwards to the south end of Hag Spring Wood, where its junction with the Upper Permian sandstone may be clearly defined, as rocks of these two formations crop out within a short distance in the same field. Continuing eastward it crosses the "Vale of Nightshade," on the north side of Little Mill, ranging between Billing Cote and Abbots Wood, thence south-south-east between Newton and Furness Abbey Park to the iron ore works of Mr. W. Boulton, on the Parkhouse estate, at which place it again joins the Upper Permian sandstone, and following the same direction skirts the brow of the hill on the east side, parallel with the Furness Railway, to the curious Crab-Rock conglomerate, near the level crossing of the railway on the Hallbeck Estate, afterwards it turns abruptly to the east and becomes interlaced with the small patch of Magnesian Limestone of Stank, which shows in a quarry on the road side between Stank and Old Hallbeck, at which place it is not more than twenty feet in thickness. A few years since, E. Wadham,

Esq., mining agent to the Duke of Buccleuch, in trying for coal, bored through this Magnesian deposit in the floor of the quarry, where it was only fourteen feet in thickness, and below it he found an immense bed of a dark coloured alum shale. It is to be regretted he did not persevere and prove what was lying beneath, as the geological conditions are favourable to the discovery of coal in this neighbourhood, and, certainly, if it does not exist here, it will be useless searching further north, as Mountain Limestone, clay-slate, or other Silurian or Cambrian rocks prevail through the whole of northern Furness, all lying in the sequence of stratification far below the coal measures. However, Mr. Wadham gave this place a very fair trial, having penetrated to a depth of nearly 300 feet, and secured the "bore-hole" for operations at a future time.

Strangers have said, either in jest or in earnest, that "it is selfish and unreasonable for the Furness people to be anxious to find coal when they have abundance of every other natural production in the kingdom, (probably there is no place on the face of the globe containing so many in so small an extent of country), and instead of being thankful to the Almighty for casting their lot in this beautiful land overflowing with everything conducive to the comfort and happiness of mankind, they ungratefully lament the absence of coal,—just like the miser whose coffers are running over with treasure,—would grasp with eagerness the last guinea in the world. Natives of the district, who are every-day observers and partakers of the blessings by which they are surrounded, are apt to underrate them, but we, who are strangers, belonging to a less favoured locality, admire and appreciate them in an extraordinary degree." This is something like the substance of a conversation we once had a few years since, with a strange gentleman whom we met in one of our geological rambles. We acknowledged there was a good deal of truth in his remarks, but we pleaded "not guilty" with respect to ourselves, not only to the charge of ingratitude, underrating, or not appreciating to their utmost extent all the blessings and comforts we had enumerated, but we were afraid we had laid ourselves open to the contrary charge of *over*rating them, for we have always looked upon Furness as the type of a beautiful little pet kingdom complete in itself, possessing not only all the elements for successful trade and commerce, but scenery that will bear favourable comparison with any part of Britain, having every variety of landscape, from the quiet pastoral beauty of Low Furness, to the wild mountain region of Seathwaite and Coniston. This favoured spot of earth, abounds with all the student of natural science can desire, either in botany, mineralogy, or geology, especially the latter, as it contains many rare and beautiful Organic Remains, many of them of unknown species. Furness must, of necessity, be beautiful, bounded as it is on one side by Windermere, that lovely "Queen of Lakes," and on the other by Wordsworth's own "Crystal Duddon," and having within it, lakes and rivers no way inferior to these waters, such as the lakes of Esthwaite and Coniston, and the rivers Brathay, Leven, and Crake, all equally pure and bright with the poet's favourites. The reader will kindly excuse this digression, for, like the mischievous boy, we could not help it, because our love for this enchanting district is so intense, that it sometimes degenerates into a sort of worship, or adoration, of the natural objects it contains, instead of giving the whole heart and mind to the worship of the Almighty Creator and Preserver of all.

The Magnesian Limestone of Stank and Old Hallbeck, is a good and durable building stone for plain work, but not very suitable for ornamentation, on account of its containing numerous small cavities like shot holes, filled with pure carbonate of lime. It is of a beautiful pale yellow colour, not absorbent or subject to discolouration by atmospheric influence, however it has not been much used here for building beyond the hamlet of Stank, but we find many good specimens of it, here and there, set up as gate-posts in the fields adjacent to the quarry. This deposit is known locally as "Tommy Berry Stone." A Commission was appointed by Government, it will be remembered, a few years since, to examine and report on all the principal quarries of building stone in the kingdom, and to select the best for the New Houses of Parliament, at Westminster. The Commissioners all agreed that Magnesian Limestone was the best for that purpose. These gentlemen acted according to the best of their judgment, and selected the Magnesian Limestone of Yorkshire, but in doing so they made a great mistake, for some of the stones forming the fronts of the Palace are, even in the comparatively mild atmosphere of Westminster, rapidly decomposing and flaking off from the surface in pieces almost as large as ordinary dinner-plates. This would not have been the case if they had taken the Magnesian Limestone of Stank for the plain and the carboniferous or Mountain Limestone of Stainton, Baycliff, or Birkrigg, for the ornamental portion of the works, besides, an agreeable variety would have resulted without producing any violent contrast of colour, and as the stone from all the places enumerated above, are almost imperishable, the British Houses of Parliament would have been as enduring as the Pyramids of Egypt. We have not succeeded in finding Organic Remains either at Old Hallbeck, or at Stank, and fossils are not abundant in the Magnesian Limestone in any part of the kingdom. From Hallbeck eastward, the course of the Mountain Limestone cannot be made out with certainty, probably it takes a north-east direction between Stank and Leece, to Dendron, where it again crops out, thence skirting the north end of the "Deep Meadows" to Gleaston, where it is almost in actual contact with a dark chocolate-coloured porphyry dyke, at the east end of the village. This strange igneous matter is semi crystalised, breaking with a splintery fracture, and has burst through the superincumbent Mountain Limestone and mineralised it by heat, to such an extent as to give it all the appearance of Chert. In the line indicated above, Mountain Limestone exists to a certainty as already shown, and although we have no rock of any kind on the surface, to the south of this line, borings for mining purposes conducted in April, 1867, proved that about a quarter of a mile south of Dendron, Upper Permian sandstone existed at eighteen feet below the surface, and that the Mountain Limestone was lost altogether, therefore it demonstrated the junction between the Mountain Limestone and the Permian sandstone, as occurring a short distance southwards of Dendron and Gleaston. From this point it probably skirts the north end of "Deep Meadows," and takes a north-east direction to the Western shore of Morecambe Bay, at or near Point of Comfort.

Besides the proof afforded by the borings alluded to, the physical features of the surface give indications confirmatory of this supposition. The Mountain Limestone extends two or three miles from this place eastwards under the sands of the bay, for not only the scars on the west shore, as Point of Comfort Scar, Leonard Scar, Newbiggin, Moat, Elbow, Church, Bien Well, Terry, Seawood, Wadhead, Coup, and Hammerside Hill Scars, are all limestone, but those also at a considerable distance from the coastline as Idridge Scar, Elwood Scar, Chapel Island, and the Black Scars are all composed of the Mountain Limestone. The southern boundary may be considered to end two or three miles east from the western shore of Morecambe Bay, and at a point opposite to Leonard Hill and Beckside, all the scars enumerated above occurring in the order in which they are given in traversing its eastern boundary northwards to Plumpton Hall, where we meet the north-east limit of the Mountain Limestone of Furness as detailed in a former chapter. We have now completed the whole circuit of this splendid deposit of Mountain Limestone, and having examined and compared it with most of the principal limestone quarries in the kingdom, we may mention that not only is there none superior, but for durability, ˙beauty of texture, colour, susceptibility of a brilliant and everlasting polish,—indeed for any purpose either economical or ornamental, — it has no equal in England.

UPPER PERMIAN SANDSTONE.

We now come to the uppermost or newest rock formation in the district, the Upper Permian Sandstone. It is a moderately good stone, but not a first-class building material, as may be seen in the venerable ruins of Furness Abbey, founded in 1127, by Stephen, Earl of Moreton and Bologne, afterwards King of England, some portions of which are rapidly mouldering away, while others are as perfect as when first erected, although some parts of the building are upwards of 700 years old. The Permian Sandstone cannot be compared for durability with Green Slate and Porphyry, Coniston Flag, Lower and Upper Ireleth Slate, and especially Mountain Limestone. In this sandstone there is a public quarry in active work at Hawcote, and two or three private ones in the same neighbourhood which have not been in use for some years. The stone in all of them is of a dark red colour, and occasionally contains small almond-like concretions of indurated clay, but no Organic Remains. This deposit is of small local extent, its northern boundary commencing on the eastern shore of the Duddon on the south side of Roanhead, ranges with the Mountain limestone south-east through the Sinkfall Estate to Mill Wood, Little Mill, and Abbots Wood, thence south-south-east by Furness Abbey and Parkhouse to Roose, where all trace of it is lost in that direction in the railway cutting on the north side of the hamlet where it appears for the last time, but in a very degenerate and shattered condition in the eastern bank of the Railway. No rock of any kind is found in Furness south of this point.

The southern boundary of the Permian formation may be taken from the east shore of the estuary of Duddon, near to Sandscale, by Ormsgill, eastward by Newbarns, to end at the hamlet of Roose. Thus it may be said the Permian Sandstone occupies a triangular space, including within its area not only the village of Hawcote, the hamlet of Newbarns, and the mansions of Abbots Wood, Mill Wood, and Crosslands, but the following farms:—Sinkfall, Brestmill Beck, Rakes Moor, Sowerby Lodge, Sowerby Hall, Furness Abbey, Parkhouse, High Cockan, and Ormsgill. It will be seen from the foregoing description of the different rock formations of Furness through the whole course of our traverse from Seathwaite and Coniston on the north, by Broughton, Kirkby, Lowick, Ulverston, Urswick, Stainton, to Hawcote and Roose on the south, we have been rising step by step in the general series of the stratifications of the earth, just like ascending a flight of stairs, the lowest step of which would represent the Skiddaw Slate, the next the Green Slate and Porphyry, then Coniston Limestone, Coniston Flag, Coniston Grit, Lower Ireleth Slate, Ireleth Limestone, Upper Ireleth Slate, Lower Ludlow Rock, the Old Red Sandstone,

Mountain Limestone, and so on until we reach the Upper Permian Sandstone and Magnesian Limestone, which is the highest step we can attain in Furness. We are not quite half way up our imaginary staircase, but if we continued in the same direction, *i.e.*, south-south-east, we should again rise gradually to higher, consequently newer rock formations, and when we arrived at the south coast of England amongst the Tertiary formation of the Isle of Wight, we should have mounted the highest step in the series, and passed over all of them. It requires a very small amount of geological knowledge to enable anyone to perceive, that, as we proceed south-south-east from Furness to the south coast of England, we are rising to higher and later depositions, and Organic Remains found in that direction assume a more modern type, so that as we approach the south-coast we find many fossils of existing species, identical with those inhabiting the neighbouring sea, and when we reach the Isle of Wight the same Organic Remains greatly predominate over those that are extinct. It is a remarkable geological fact that all the different rock formations enumerated above may be found *in situ* within ten miles of Ulverston, and it is still more astonishing when we consider that in taking a natural section from the north side of Bootle in a westerly direction to the sea coast, from Seaton Hall and High Kinmont to Selker and Tarn the whole of them are wanting, with the exception of the Upper Permian Sandstone, which lies immediately on the granite of Eskdale without any other intercalating formation.

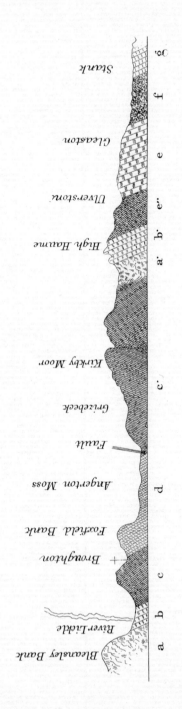

Section from Duddon Bridge by Ulverston to Stank.

a a′ Green Slate and Porphyry.
b b′ Coniston Limestone.
c c′ c″ Coniston Flag.
d Coniston Grit.
e Mountain Limestone.
f Upper Permian Sandstone.
g Magnesian Limestone.

ITINERARY.

A minute anatomical acquaintance with the bones and muscles, is deemed essential to the painter who grapples with the difficulties of the human figure. Perhaps when the geological vocabulary shall become better incorporated than at present with the language of our common literature, a similar acquaintance with the stony science will be found scarcely less necessary to the writer who describes natural scenery. Geology forms the true anatomy—the genuine osteology of landscape; and a correct representation of the geological skeleton of a locality will be yet regarded, I doubt not, as the true mode of imparting adequate ideas of its characteristic outlines.—Hugh Miller's First Impressions of England and its People.

DUDDON BRIDGE TO ULVERSTON, HAWCOTE, AND STANK.

WE have no doubt it will interest many of our readers to know the particular localities of all the different stratifications of Furness existing near Ulverston, where each may be seen and examined *in situ*. For the purpose of illustration we will suppose a traverse from Duddon Bridge to Ulverston, Furness Abbey, and Stank, imagining all the superficial "drift" and other material away, and the rock laid bare through the whole course of our route, we should then have a sequence of the following rock formations, beginning with the lowest and ascending gradually to the highest in Furness. Our course would be as follows:—
On the north side of Duddon Bridge at Bleansley-bank and

the Heights, there crops out boldly a hard light-green-coloured slaty rock, the Green Slate and Porphyry, the lowest and oldest rock formation in Furness, and with the exception of the Skiddaw Slate, the oldest stratified rock in the north of England. Bleansley-bank is the southern limit of the Green Slate formation, but it extends northwards beyond the northern boundary of Furness, and several miles into the counties of Cumberland and Westmorland, without any change whatever. After leaving Duddon Bridge, a little way on our road to Broughton we pass over a narrow band, not more than one hundred yards in breadth, of a dark blue rock, ranging up the valley of the Lickle, in a north-east direction, the well known Coniston Limestone, the equivalent of the "Bala Limestone" in Wales, and it continues in this direction upwards of fifty miles. It is eminently fossiliferous, yet where it is sound and retains its natural blue slaty colour, very few fossils can be found, although the rock abounds with them. However the Coniston Limestone is not enduring but weathers and decomposes rather quickly when exposed to the influence of the atmosphere.

We cannot cross this highly interesting deposit without a few words. The Coniston Limestone may be considered as the lowest and first great store-house of animal life in the rocks of the earth, and contains many rare species of beautiful fossils of high organisation, thereby affording strong evidence against the "development theory" of Darwin and Lamask, which we have always considered nonsense, but it is not necessary to give our reasons here. Certainly we do not deny that the Skiddaw Slate and its equivalent rocks in Wales and Canada, are sparingly fossiliferous also, but the Coniston Limestone is the lowest stratification that abounds with them, and the first where fossils may be obtained in quantity. The Coniston and Bala Limestones are known to all geologists in

Europe and America. After crossing the river Lickle, and before we ascend the brow of High Cross hill, leading into Broughton, we enter on the Coniston Flag, a beautiful slate rock of a dark blue colour, ranging in a north-east and south-west direction. In some parts of its course it possesses perfect slaty cleavage, and in all it is a most excellent building material. Broughton, Torver, and a great portion of Coniston stand upon it. This formation, like the Green Slate and Porphyry and Coniston Limestone, not only extends from the point where we cross them forty or fifty miles in a north-east direction, but also several miles to the south-west into Cumberland. None of the others, with the exception of the Mountain Limestone, extends into Cumberland, all being cut off by the estuary of Duddon, and it is not quite certain even that the Mountain Limestone crosses the estuary, forming, we suppose, a sort of horn projecting from the great Mountain Limestone deposit of Furness.

Immediately after leaving Broughton, on the road to Ulverston, we come to the Coniston Grit, which, ranging with the others, occupies the whole of the rough rocky land south-west from the Ulverston Road to the Railway Station, at Foxfield, where it is cut off and entirely disappears. But in the contrary or north-east direction, like the formations just mentioned, it extends upwards of forty miles, crossing the lakes of Coniston and Windermere into Westmorland. We now approach Wreaks Causeway, which crosses the great level tract of land called Angerton Moss, and if all the peat-earth and other superficial matters were swept away, we should see the junction of the Coniston Grit and the Lower Ireleth Slate (Coniston Flag). These two important formations certainly join under Angerton, for we follow the red-tinted Coniston Grit down the moss on the Broughton side, and there rises up the blue rock of of Lower Ireleth Slate as soon as we reach the hard land on the east or Kirkby side, which continues through Grizebeck and a considerable way up the brow of Kirkby Moor, where it gradually changes and improves until we come to the quarries at Gawthwaite, where it possesses perfect slaty cleavage, and is in fact the Upper Ireleth Slate. This cleavage continues through these quarries in a north-east direction to Stone Dykes, Lowick Low Common, Parkgate, Lowick High Common, and Groffacrag Scars, after which it gradually disappears until by Iron Yeats and Knott-hollow, on the road towards Ulverston, scarcely any part of the slate is suitable even for common flags, and at Gamswell, Hoad, and Flan Hill, it has degenerated into coarse walling stones of very little value.

In the beck-course, in Gill-banks, on the north side of the town of Ulverston, a small patch of Old Red Sandstone crops out, and immediately overlying it fragments of Mountain Limestone, but they are utterly worthless. Ulverston stands on a deposit of Diluvial Drift, the railway cutting on the south side of the town, showing a depth of forty or fifty feet of that material, without coming to the lowest of the deposit, and about half a mile further on, in the bottom of the beck in front of Swarthdale House (G. Sunderland, Esq.,) the Mountain Limestone is seen in mass. This is the northern boundary of the great Mountain Limestone deposit of Furness, which continues for upwards of six miles in a southerly direction to Gleaston and Dendron, where it is lost under the "Deep Meadows," but we will now follow it from Ulverston in a south-westerly direction to Lindal and Dalton. About half a mile south of the latter place we come upon the Upper Permian Sandstone of Hawcote and Furness Abbey, and at Stank and Old Hallbeck the Magnesian Limestone,—beyond this southwards there is no rock seen in Furness. It will thus be seen that within a radius of ten miles, with Ulverston as a centre, we have eleven different stratifications *in situ*, in one natural section, and without deviating materially from a straight line.

PLEASURES OF GEOLOGICAL PURSUITS.

HAVING described the boundary of the district under consideration, and taken traverses in the line of strike of all the different stratifications enumerated, as well as a section of them minutely described where each may be studied *in situ*, and as the geological student cannot make much progress in the parlour with books alone, it is necessary he should have considerable experience in the field, and possess at least an elementary collection of Organic Remains, so we will now point out several different localities favourable for obtaining them. In following out the above proposition we will commence with the highest fossiliferous stratification in Furness, and after rambling to several likely places and securing all we think requisite for our purpose there, we will descend in the series of deposit to the next, and thus visit the whole in rotation, making excursions to all places of more than ordinary geological interest. For the most part these must be travelled on foot, and the student will enjoy many delightful strolls. We cannot here refrain from saying a few words on the pleasures of geological pursuits in comparison with other field studies. "Field Clubs" are become very popular, and deservedly so, for they are outdoor schools where real learning may be acquired with advantage, and we recommend all young people to join them. Botanical excursions are both pleasant and instructive, and it is no doubt very gratifying to fall in with a rare and curious plant which, although rare, may not be new, thousands of people having seen it before, moreover, it may have been described and published in botanical works for many years — how much more delightful then will it be to the geologist when breaking up a portion of fossiliferous rock, Coniston Limestone for instance, with every blow of the hammer he expects to develop not only a new species, but new genera also, hitherto unseen by man, and absolutely unknown to science.

In the above statement there is nothing fanciful or unlikely, but a high probability that he will succeed in finding something new in the course of a few hours, and if he be really smitten with the geological fever, he will soon become entirely absorbed in the pursuit, for such at any rate has been the result of our own experience. Many times, dreamy abstractions, though always of a pleasurable kind, have drowned sober thought and allowed the mind to wander back through all time, when night coming on has found us several miles from our intended home. In such a case we have laid down in some quiet and sheltered part of the rock, and slept until morning, dreaming perhaps of the strange Organic forms we had been contemplating through the day and speculating on the condition of the earth at the time these fossils had life and motion, thousands and thousands of generations before the creation of man. Geology has a peculiar tendency to induce the mind to reason upon long cycles of the past, forgetting the present and the future, sometimes to such an extent that when we have been awakened in the night by the croaking of the raven or a sudden blast of wind, we have remained some time before we could recollect in what part of the world we laid down to rest. To be placed in a situation like this, entails a very small amount of physical discomfort, which any man in health may endure for a few nights during seven months in the year, and there is a grandeur, truly poetic and sublime in the thought that he is shut out, as it were, from all the world, perhaps thousands of feet above the habitations of man, and there almost in the actual, visible presence of the Almighty, communing with his own thoughts in all seriousness and humility. Nay more, it is only under circumstances like these that man can thoroughly appreciate all the beauty and grandeur of the natural phenomena by which he is surrounded; either in the storm and tempest, which almost shakes the mountain under his feet, or in the calm and peaceful moonlight of harvest time when the wind is softly waving the corn many hundred feet below, and with a gentle hush sighing amongst the heath and scanty herbage above his head, till at last, even the murmur of the wind ceases, and a deep, holy silence pervades all nature, leaving him entirely and absolutely alone with his meditations, but still under the protection of the Almighty, to rest through the night in quiet and safety. We have experienced both these conditions of nature several times, and the remembrance of them gives a pleasure we cannot describe. Again, to be caught up on a mountain by the approach of night is no great hardship to any man, but it belongs almost exclusively to the enthusiastic geologist to realise it — the Waltonian keeps to the rivers and streams below, and the botanist does not ascend so very high the mountain side, but the geologist goes to every place where rock can be found. The rocks of the earth belong to him almost exclusively, and he exercises acts of ownership over them accordingly. These have been some of the pleasantest days of our life, and although we have seen seventy-nine birthdays, we are truly thankful to be blessed with health and strength for the work, even to this day, and we hope to have many more quiet and comfortable nights, free of cost at the "Rock Hotel," for there is a peculiar and awful solemnity in the thought of sleeping alone on the mountains, which the inexperienced cannot understand. We might enlarge on this subject, but it would be out of place here, for although it has afforded us a pleasure, we do not advance it as one of the chief gratifications of geology, but as a proof of its influence on the human mind, and through sympathy to enable the body to endure more than an ordinary amount of hardship and fatigue. We conclude these remarks with a quotation from the Rev. W. G. Barret's Lectures on Geology:—"In marking the characteristic fossils of each formation, let us suggest in passing the vast amount of pleasure there is in going to a friend's house and looking at the minerals, or Organic Remains, that may be in the cabinet or on the mantle-shelf, and being able to take them up one by one, and to say this is from the Silurian; that is from the carboniferous; this is from the cretacious; and that from the Wealdon formations, and so on. Why, it gives a magical feeling of delightful interest to every object we see, and will always make a person a welcome visitor with friends with whom, instead of talking scandal he can talk geology."

FROM ULVERSTON TO PLUMPTON AND THE SEA-SHORE.

We will suppose the student or amateur, provided with the outfit described at p. 54. Starting from the centre of the town, by way of Church Walk, he will pass the beautiful church of St. Mary, one of the finest churches in Britain of its dimensions, the restoration and beautifying of which was mainly due to the exertions of the Rev. Canon Gwillym, M.A., the late lamented incumbent. Several other kind hearts contributed liberally to the pious object, but the rev. gentleman made it the second object of his earthly thoughts, the first being to visit the poor and needy, to feed the hungry and clothe the naked. With ample means of his own apart from the living, of which he never received a penny in his life, he relieved the wants of all truly deserving, and when the Lord in His own good time called him hence, he was mourned by hundreds whom he had relieved and comforted. He departed with their blessings, and his grave was watered with the tears of a vast crowd of sorrowing parishioners.*

After passing through the church-yard we skirt round the foot of the Hill of Hoad, the southernmost hill of Silurian clay-slate in Furness, but we must let it alone at present, to take a few rambles amongst the carboniferous or Mountain Limestone, only we may just notice in passing, the extraordinary upheaval of this Silurian hill, which although deposited almost horizontally, has been so tilted up by some great convulsion of nature that the stratification or true bedding of the rock forms in part of its range an angle of 78° with the plane of the horizon. Organic Remains are exceedingly scarce in the quarries of Hoad, but we have obtained specimens of two different species of undescribed corals, and a very good trilobite, *Phacops lævis*, with one of its eyes well developed, now in the cabinet of Alexander Brogden, Esq., M.P. No attempt should be made to develop any good fossil until brought home and

* Any stranger after enjoying the enchanting view from the summit of Hoad, and examining the interior of St. Mary's Church, would naturally wish to have seen the incumbent, who has been represented as a true Christian minister, labouring in his Master's cause with praiseworthy zeal and industry. He would rarely have been found at home, but generally in the streets, especially of the poorest parts of the town, and his slender, bent form might have been recognised daily, walking very quickly, carrying a small basket containing delicacies and comforts for some of his sick poor, and if he met any old claimants for his bounty on the way, his purse was out in a moment. The above is not a fanciful sketch, but almost literally true, and the acknowledged character of the late much esteemed vicar, of whom we are proud to record this tribute, as being justly due to departed worth.

"To relieve the wretched was his pride,
And e'en his failings leaned to virtue's side."— *Cowper*.

more perfect appliances can be brought to bear upon it, such as small chisels made from spindles, hackle-pins, darning needles, &c. An assortment of sewing needles from the coarsest, to the finest number made, will be found very useful. Proceeding down the north bank of the Ulverston Canal, we come to the gunpowder slip at or near which place the junction occurs between the clay-slate and the carboniferous or Mountain Limestone, and extends in a southerly direction for upwards of six miles without any interruption whatever. Half a mile beyond the slip we come to the dark limestone quarry of Plumpton, where we shall find several species of fossil shells, principally *spirifer*, *orthis*, and *producta*, and some good specimens of the genus *Cornulite*, especially the *Turbinola fungites*, but most of them are firmly imbedded in the rocky floor of the quarry and cannot easily be detached. We have no doubt several loose ones may be picked up, all showing interior organisation and susceptible of a high and lasting polish.

In the old workings of this place, where the rock is weathered by long exposure to the atmosphere, there are developed numerous cross sections of several species of the genus *Cornulite*, all of them showing interior organisation, some their Organic structure parted into flakes like the cross section of the genus *salmo* and other recent fish; portions of the columns and plates of the heads of two or three different species of Encrinites are also exposed, their markings and sculpturing being so rough and obscure, that they are worthless and may be let alone, as we shall find beauties at another place. Organic Remains are not abundant, and with the exception of a few specimens of *Producta gigantea* to be obtained, nothing very interesting has ever been found at this quarry. A short distance eastward to the sea-shore in another quarry of Mountain Limestone, the student will be surprised to find the rock is not dark like that he has just left, but a beautiful white, similar to that of Baycliff, Birkrigg, and Stainton, although the distance from one quarry to the other is not more than two hundred yards. There we have also a lime-kiln almost in constant work, which produces lime of a very good quality both for building and agricultural purposes, but the dark stone of the other quarry is preferred for buildings of a good description, and when erected in the class of work technically called " bedded and jointed," is exceedingly beautiful and durable.

We now go northward between Plumpton Wood and the shore of the bay, and near the Red Hole (an old trial mine for iron ore), there occurs a beautiful floor of limestone rock, washed by the sea, and when left wet by the receding tide, numerous species of Organic Remains are exposed firmly imbedded, many showing interior organisation and displaying their original colour unfaded by time, which is a very pretty and a rare circumstance indeed, and with the exception of the smooth floor of rock on the sea-shore, near Baycliff, is the only place in Furness where this phenomenon occurs.

Besides the various species of coloured *producta* and other large fossil shells here, there are numerous "sops" or detached masses of coral entire, consisting of from twenty to fifty stems or polyps, every specimen perfect in itself, and unconnected with any other, so that if quarried either by blasting or other suitable means, and polished, leaving a border of plain ground all round, it would show the fossil distinct and clear, and be exceedingly beautiful, as each specimen would be complete and not a portion broken from a larger mass of the same. We are at a point on the sea-shore about 200 yards from Plumpton Hall, very near the junction between the clay-slate and the great deposit of Mountain Limestone, which extends entirely across this part of Furness, from the western shore of Morecambe Bay, to the eastern shore of the estuary of Duddon. From

Plumpton Hall northwards for a distance of five or six hundred yards, there is no rock on the shore until we come to the splendid viaduct across the western portion of Morecambe Bay, where we find the slate rock in considerable force, being that member of the slate series which Professor Sedgwick calls Lower Ludlow Rock, a mass of it forming the western abutment of the Leven Viaduct of the Furness Railways. North of this spot the Lower Ludlow Rock extends on the sea-shore about half a mile, when it is interrupted by a small patch or island of Mountain Limestone, covering only an area of eight or ten acres and forming the extreme northern point of this rock on the western shore of the bay. There is a small quarry here yielding stone of an exceedingly fine grain or texture, locally called "Bloomery," used at the furnaces of Newland and Backbarrow for smelting iron ore, the only furnaces in Lancashire, or rather in England, where iron ore is smelted with wood charcoal, with the exception of one at Duddon Bridge, in Cumberland, belonging to the same firm (Messrs. Harrison, Ainslie, & Co.) Very few Organic Remains have been obtained at this place, but the slate rock by which it is surrounded is sparingly fossiliferous, two or three imperfect specimens of *Cardiola interrupta* have been found, as well as a few of *Pterinea*.

ULVERSTON TO BIRKRIGG, BAYCLIFF, AND THE SEA-SHORE.

IMMEDIATELY on leaving the town of Ulverston in a southerly direction, we come to the deep cutting at the station on the Furness Railways, the sides exposing a most interesting section of gravel, boulders, and quartz sand, with three different beds of sea sand intercalated, all the different members preserving their true relative positions in the series of deposit, and subject to the same waves and contortions with the gravel and other matter with which they are associated, the whole in a general way conforming to the swells and depressions of the land through which they pass. This section continues westward into Pennington, a distance of more than a mile, and the depth of the cutting increases from about forty feet at the Ulverston Station to sixty or seventy feet in Pennington. The immense mass of "diluvial drift" is altogether of Silurian origin, and almost entirely without Organic Remains. After several careful examinations we have succeeded in finding one decomposed boulder containing several minute fossils, all of them Silurian in type, which interesting specimen is now in Mr. Brogden's cabinet. The sea sand is undoubtedly entirely unfossiliferous, and we have also obtained specimens of three different porphyry dykes. It is highly probable this immense mass of "diluvial drift" was deposited by a wave of translation from the north-west, especially as it contains no fragments of limestone. In this part of the deep cutting no rock of any kind is seen, but theoretically it may be assumed as the junction between the uppermost member of the clay-slate formation and the great deposit of Mountain Limestone of Low Furness, as the limestone rock crops out in the bottom of the brook about 400 yards further to the south, where it may be seen from Levy Beck Bridge. After passing Swarthdale House, proceed by Mountbarrow to the "Red Lane," or old Roman road,—the first, and for a considerable period the only road in Furness,—in repairing which about fifty years since, a portion of tesselated pavement was laid bare, but no record exists to show whether it was destroyed or covered up again. In some newly erected walls along the side of the road, many interesting Organic Remains may be seen, particularly in the stone coping. We next come to the limestone quarries on the west side of Birkrigg, where Organic Remains are not abundant, or of much interest, consisting of two or three species of *producta*, two or three of *spirifer*, and a few of *orthis*, none of which can be clearly developed. We will now ascend to a small quarry on the summit of the hill, which is not much better than the one below; *Cornulites* of two or three different species are plentiful, but all so firmly imbedded in the face of the rock that they cannot be extracted without much difficulty, the most interesting fossil, and one apparently peculiar to the place, yet even scarce here, being an undescribed species, somewhat resembling the *Turbinola fungites*, except that it is surrounded by a punctured border which, when polished, is exceedingly beautiful. After leaving, walk down the slope of the hill, to a small trial quarry five or six hundred yards eastward, which abounds with fossils, principally *spirifer* and *producta*, the latter especially plentiful, and although generally mutilated and imperfect, some good specimens may be obtained.

Half a mile further eastward to the bottom of the hill, in a small private quarry behind the pleasure-grounds of Well Wood House (Mrs. Petty), numerous good specimens of the *Producta gigantea* may be obtained, some large enough when living to contain a shellfish two pounds in weight. Before leaving Birkrigg, we ought to have visited some quarries on the eastern slope of the hill, which have yielded good specimens of the *Eumphales crustatus*, and where, here and there, sops or pockets of copper ore of a good quality occur, while in the adjoining wood (Seawood), there is a small copper mine carried on by a spirited little company, whose works are not yet fully developed, although it may be a profitable concern in a little while. At the excavations on the shore adjoining Seawood, where limestone rock has been quarried for exportation, Organic Remains are rather scarce, the only fossil of interest being *Edmunda sulcata*, a type of the edible mussel, some exceedingly good specimens having been obtained. A short distance southward is the village of Baycliff, where two public quarries are in work, both producing almost pure white stone, in blocks of almost any dimension, sound, and free from cracks and "drys," susceptible of a brilliant and enduring polish, every way suitable for ornamental sculpture, with a fine "arridge," and not breaking with a splintery fracture.

We have compared the stone of Baycliff with limestone from the principal quarries in England, and with the exception of the grey dappled marble of Stainton, two miles west of Baycliff, we believe it has no equal in the kingdom. Organic Remains are rather scarce at both these quarries, and from the compact and solid texture of the stone, the few it contains are very difficult to develop, and consist principally of our old friends *Producta gigantea* and *Spirifera*, beside which at the northern quarry a few exceedingly good specimens of *Orthoceratite* have been obtained with beautiful interior organisation, one gigantic specimen being nearly four feet in length and a sufficient load for the quarryman to carry home. The stone of the southern quarry is similar and in every way equal to that of the northern, and Organic Remains are rather scarce here also, but we have obtained a splendid specimen of an undescribed species of the genus *Orthoceratite* upwards of fifty pounds in weight, nine inches in diameter, only two diameters in length, and very much resembling a Swede turnip, the lines of structure on the outside perfect, showing portions of interior organisation. We have only seen one other specimen of the same species, and that was very imperfect, just showing sufficient to determine it to be of the same species, at the large limestone quarries at

Scrimerstone, near Berwick-on-Tweed. We also found at Scrimerstone a portion of a most gigantic *Orthoceratite*, which, when whole, must have been at least eight or ten feet in length and upwards of half a ton in weight, as the portion alluded to was as thick as the body of an ordinary man, and although only three feet in length was as much as two men could lift from the ground, so that it may perhaps be lying there at this day. We also saw in the pleasure-grounds of the owner of the quarry, at Sea House, two other large *Orthoceratites* of the same species. It is wonderful to contemplate these monstrous chambered shells of the carboniferous period. The extensive quarries at Scrimerstone, are situated on the shore of the German Ocean, three miles south of Berwick-on-Tweed, where the great deposit of Mountain Limestone extends inland in a westerly direction to the villages of Lowick and Oxford, where large quantities of the stone is constantly worked, and where Professor Sedgwick obtained some good specimens, figured in his splendid work on the Geology of the Lake District. From the village of Oxford it stretches in a north-westerly direction, by Ford, Palingsburn, and Flodden Field,* and continues in that direction several mile beyond Coldstream. We have examined this immense deposit through the whole of its range, which is much greater than that of Furness, and we have purposely

---

* Flodden Field the scene of a most important and decisive battle, on the 9th September, 1513, between the English forces under the Earl of Surrey, and a splendid Scottish army commanded by King James IV. in person, the battle was maintained with the utmost determination and bravery on both sides, but was most disastrous in its consequences to Scotland. King James was killed on the field, and when his body was found the morning after, there were stretched in death around it, a considerable number of the principal nobles of his kingdom, showing that part had been the hottest in the fight. It was then uninclosed, but is now a highly cultivated field, and on the identical spot an immense undressed block of Mountain Limestone, several tons in weight is erected, called the "King's Stone." This battle, so fatal to Scotland, and in which almost every great family in that kingdom lost a member, gave rise to that touching and pathetic Scottish lament, "The Flowers of the Forest."

digressed in this ramble and taken a flight of 150 miles northwards, from Baycliff to Berwick-on-Tweed, to compare it with that of our own native rock, and we confess it gave us a sort of selfish pleasure to find that in every part of this great range of country, their Mountain limestone made a very unfavourable comparison with that of Furness. Certainly, when burnt, it makes good lime, but as a building stone it is almost worthless. We must also further digress to state that when residing in that district, we took a stroll to Holy Island, a continuation of the same deposit eastward, into the German Ocean, expecting from Sir Walter Scott's "Marmion," to find some interesting Organic Remains, but we were disappointed.

> But fain Saint Hilda's nuns would learn
> If on a rock by Lindisfarn,
> Saint Cuthbert sits, and toils to frame
> The sea born beads that bear his name:
> Such tales had Whitby's fishers told,
> And said they might his shape behold,
> And hear his anvil sound;
> A deadened clang—a huge dim form
> Seen but, and heard, when gathering storm,
> And night were closing round.—*Marmion, Canto Second.*

Sir Walter's poetic notice here quoted has induced every stranger who visits the island, to obtain a few "Cuthbert Beads" at any price, although they are both scarce and worthless.

The rock alluded to above is situated near the south west shore of the island, close by which we saw several children crawling amongst the shingle, seeking for the "beads," and we laid down amongst them to find a few, but they were so utterly valueless, that we thought it would not be a bad speculation to take a thousand or two of our own beautiful Cuthbert Beads, and a few sculptured plates of the heads of the Encrinite, from Hawkfield, as they would more than pay for the journey. There is an extensive limestone quarry on the island always in active work, which we examined minutely, but could not find any trace of Organic Remains. The stone produced at this quarry is burnt on the island, and makes good lime, which is principally shipped off to distant places, but the stone like that of Scrimerston, Lowick, and Oxford, for building purposes is almost useless, and so far from rivalling that of Baycliff and Stainton, is not equal to the worst limestone quarry in Low Furness.

ULVERSTON TO SWARTHMOOR, URSWICK, HAWKFIELD, AND GLEASTON CASTLE.

RAMBLING again southward by Levy Beck and Swarthmoor, we pass the quaint little Meeting House, presented by George Fox to the Society of Friends, of which religious body he was the founder. The date of its erection is indicated by an inscription over the entrance—"Ex. Dono. G. F., 1688." A small congregation of Friends still assembles in it, and the building, externally as well as internally, presents a simple appearance, characteristic of their mode of worship. Half a mile west from this place is Swarthmoor Hall, the ancient home of the Fells of Swarthmoor, where the noble and amiable Margaret Fell, before and after her marriage with George Fox held meetings for worship, for which, and refusing to take the oath of allegiance, the former was imprisoned in Lancaster Castle, afterwards appearing before the judges who were utterly confounded by her defence. George Fox also suffered much persecution, imprisonment, and other hardships, an account of which is given in his Journal, and in a most interesting work by Maria Webb, entitled "The Fells of Swarthmoor." There are several other interesting

URSWICK TARN.

historical reminiscences connected with Swarthmoor, but as they might be considered out of place in a geological ramble, we will pass on by Trinklet to Much Urswick, a distance of two miles, along which route there is nothing of geological interest. The village of Much Urswick is situated in a deep basin nearly surrounded by high ground, and encloses a tarn, the water-way covering an area of considerable extent, to which there has always been attached a vast amount of mystery. We have given some of the legends respecting the origin of the tarn in the "North Lonsdale Magazine" for September, 1866, and although these stories are of a wild and romantic character and utterly unfit for the present generation, they were almost universally believed seventy or eighty years ago. We also gave in the October number of the same periodical, a short article on the tarn, the principal part of which we here reproduce. We do not presume to account for the origin of the tarn, but to give the result of some experiments which have a direct bearing on its physical history at the present time, and to prove its depth by "soundings" when a favourable opportunity should offer for that purpose.

Much Urswick Tarn, even when divested of its accompaniments of mystery and romance, may still be considered a place of more than ordinary interest; its physical history differs materially from that of any lake or tarn in the north of England, and affords a valuable lesson to the naturalist, but more especially to the student in geology.

The tarn of Much Urswich is situated in a deep "basin," three miles south of Ulverston, bounded on the north and west by the village; and open to the south, whence its outlet takes the same direction, and after a junction with the stream from Mere tarn and the splendid springs of Gleaston Castle, passes through the village of Gleaston, and falls into Morecambe Bay at Leonard Hill, near Beckside.

In form it is an elongated oval, 412 yards in length, and 200 yards in breadth. It is the property of the Duke of Devonshire, and covers an area of 14A. 1R. 12P. The tarn is almost completely encircled by a thick belt of reeds *(Phragmites)*, flags *(Iris)*, bullrushes *(Cyperaceæ)*, &c., except at the " coot-stones," where is an open space ten or fifteen yards in width, and the north-west portion of the tarn for about half an acre is flagged on the surface of the water with broad green circular leaves, and the beautiful pearl white cups of the water lily *(Nymphea alba)*. There is only one inlet or "feeder" to the tarn, viz., the "Clerk's Beck." This inlet is forming an interesting delta of fine red mud where it enters the tarn, and illustrates in a beautiful manner (on a small scale) the formation of the great deltas of the Ganges, Mississippi, Nile, &c.

Having been engaged for several months under the Commissioners for the Commutation of Tithes in England and Wales, in a survey of the parish of Urswick, and having had a strong desire for many years to know more of this mysterious tarn, we established fixed marks, or "stations," all round the tarn, which could be referred to at any time, so that we should be enabled not only to take soundings in any part we desired, but also to put the depths of the several soundings on the map of the tarn in their true relative positions.

In the winter of 1852-3, several days and nights of hard frost occurring in succession, the tarn was covered entirely with clear smooth ice, so that the boys of Ulverston and Urswick quickly brought out their skates, but some of them required new straps, others wanted grinding, or else the wooden skate soles were entirely worm-eaten, and many a father and mother will remember how humbly they were entreated to buy their sons new patent skates.

At this time, George Kirkby, a mysterious little joiner and skate grinder, and one who dealt largely in romance, had to work almost night and day, repairing skates. We were also busy preparing apparatus for sounding the tarn, and while doing so had to visit his shop once or twice for that purpose. In a conversation with George about the tarn, he said,—" Ye'll find it a queer spot, for I assuer ye thair is weed in Girt Ossik Tarn beath thicker an' langer than t' biggest tree i' Bardsa Park." However, we became furnished with a long nautical sounding weight, one inch in diameter, hollowed out at the bottom into a sort of cup, and filled with tallow, to show what sort of material was at the bottom of the tarn; also a brace and a nine-eighth brace-bit, to bore holes through the ice; and that we might not be short of sounding line, we provided 2,000 feet of very fine and strong whipcord, and furnished with these and our "field-book" for registering the depths of the different places sounded, we proceeded to Urswick, where we found several young men and boys displaying their graceful evolutions on the tarn, before the ice was perfectly safe for skating. As soon as we made our appearance on the ice for the purpose of boring it with the brace and bit, although we assured them it would not spoil their skating, as the holes would freeze over in an hour, yet these thoughtless young men clustered round us like bees, with the intent to break the ice, but we had the presence of mind to move quickly off the deep water, or the consequence might have been lamentable; however, nothing serious happened, yet one forward young gentleman will long remember the wetting he got on that occasion.

After this mischance, we waited three or four days, the frost still continuing, the ice became perfectly safe, and we were allowed to proceed with the soundings without opposition, although the same young men were again exhibiting their skill on the ice.

Our first sounding operation was to establish a straight line across the tarn, about 80 yards from the north end, and from a fixed point on the west side to another on the east side, to bore holes through the ice at equal distances one from another, registering the depth at every place, then to move southward about 80 yards more, and from a fixed point on the west side, produce a line to a station on the east, and bore four holes at equal distances as before; afterwards a third and a fourth line, so as to have sixteen places for soundings, by which operation the tarn was divided into twenty nearly equal portions.

We then tried the depth of the water at No. 1 hole, with the 2,000 feet line, but we were rather disappointed when the lead stopped at 32 feet in soft mud, and after being assured it was at the true bottom (by lifting the lead and letting it fall again suddenly), we then raised the weight to the surface, and found the tallow in the cup at the bottom of the lead was covered with soft *red* mud. We were rather surprised at this, because the hole was about 100 yards from the inlet to the tarn at Clerk's Beck, which proved that the red water brought down from the mines at Lindal, by this stream, pervaded the whole tarn, and after being held in suspension for a while, subsided in the condition of soft red mud. We then sounded at the second and third holes, which were not much deeper, the fourth being only 29 feet deep, the lead in every instance proving that soft red mud was at the bottom.

It is not necessary to give in detail each separate measurement; but to state generally that the result of the experiments proved the tarn deepened gradually towards the south end, the deepest sounding being 41 feet; and that the bottom consisted of soft red mud, equally tinged with colour in every part of its whole area. From the above data we may assume the average depth of the tarn at Much Urswick does not exceed 39 feet; that it is gradually filling up; that a time will come when there will be no tarn there; that Clerk's Beck will run through a meadow, which now occupies the site of the present tarn, to the outlet at its south end; and that the substratum under this new meadow, to a great depth, will consist entirely of impure iron ore, similar to that of Mr. Archibald, from Nova Scotia, containing probably about 28 or 30 per cent. of iron. In the above statement we have not propounded any original fanciful theory, but a mathematical truth which can be demonstrated.

Taking the average depth of the water to be 39 feet (which is something more than we have proved it to be), and that it occupies an area of 14A. 1R. 12P., gives 24,335,883 cubic feet of water in the tarn at the present time, or rather at the time of the experiment (1853). Now all the water brought into the tarn by Clerk's Beck, its only inlet, is highly charged with iron ore and other solid matter, partly as a chemical solution, but principally as a mechanical mixture, which, after a time subsides; and as nothing but clear water issues from it at the south end, its only outlet, this solid matter is accumulating at the bottom, and will ultimately fill it up entirely.

The tarn of Much Urswick, as before stated, has but one inlet, namely, Clerk's Beck, which now forms a part of the "water level" from the mines at Lindal Cote to Much Urswick, taking also the drainage of all the mines near Lindal, and of the whole watershed of the Lindal valley.

From what has been stated, it is certain that the tarn is *now* filling up, yet the data for calculating the rate at which this process is going on, are very imperfect, as certain facts have to be assumed which are not susceptible of proof. First there must be supposed an average velocity and an average sectional area for the stream of Clerk's Beck, also the average amount of solid matter contained in the water during the whole year—none of which conditions is constant, but all vary according to circumstances. Now it is not a difficult matter to solve all these problems for any particular day, but it would require a long series of experiments even to approximate to the true quantities for a whole year. However, we will give the result of some observations and experiments bearing on the subject.

We have found the velocity of the stream of Clerk's Beck at one time = 140 feet per minute and its sectional area 2 feet; giving 280 cubic feet of water in a minute. At another time, the velocity 30 feet, and sectional area 2 feet, thus giving only 60 cubic feet a minute; but even these numbers do not truly represent the greatest extremes between the highest and the lowest velocities, and the highest and lowest sectional areas of the water in Clerk's Beck, there being exceptional times giving both higher and lower velocities. We will take the mean of these numbers for data, and that we may not exceed the true quantity, we will only take 200 days in the year, which is equivalent to giving a slower mean velocity of the stream, and also a less sectional area, thus leaving a considerable margin in favour of a slower filling up of the tarn than is due to the case.

Reckoned in this way, the quantity of water flowing into the tarn at Much Urswick will be as under, viz.:

In one minute . . . . . . 170 cubic feet.
In one hour . . . . 10,200 ,,
In one day . . . . . . 244,800 ,,
In one year, or 200 days. 48,960,000 ,,

To ascertain the quantity of solid matter contained in one cubic foot of the red muddy water flowing into the tarn (and consequently the principal element for deducing the whole amount), we instituted the following experiments:—We filled a graduated glass tube with the water that issues from the south end of the tarn, and weighed it very carefully, which by calculation gave 1·022 ounces to the cubic foot; we then evaporated all the moisture from the tube and filled it with water from the Clerk's Beck, which we also carefully weighed, and it gave 1·032½ ounces to the cubic foot; the difference between these results (minus the bulk or volume of solid matter) gives the weight of solid matter contained in one cubic foot of water from Clerk's Beck. We then took mud from the tarn, at the inlet, at the foot of Clerk's Beck, which after being partially dried, and subjected to a high degree of pressure, gave a specific gravity = 2·33, therefore, each cubic foot of the water contained 4½ cubic inches of solid matter, or, in round numbers, about 1 in 400.

From the above calculations and experiments it is deduced (without giving the mathematical formula) that Urswick Tarn is filling up at the rate of about half an inch annually, and as it had an average depth of 39 feet in 1853, it should in the winter of 1866 have an average depth of 38 feet 5½ inches; and if all the conditions remain the same, after a lapse of 923 years, Urswick Tarn will be filled up solid, and a fine rich meadow will occupy its site, with Clerk's Beck running through the middle of it. The rate at which this is being accomplished cannot be given with mathematical truth, yet the above may be taken as an approximation to it. It would be desirable to verify the soundings of 1853, when the tarn is again in a condition favourable for it; and as several of the identical places can be found from our sounding notes, it would be no difficult matter to prove whether the rate of deposit is greater or less than what is here represented.

The botanical history of the tarn is deserving of some notice; for besides the species before mentioned there are several others, such as the water violet *(Hottonia palustris)*, yellow water lily *(Nuphar luteum,)* water crowfoot *(Ranunculus aquatilis)*, &c.; and in the outlet brook are found splendid specimens of the *Anodon cygneus*, or large swan mussel, some of them weighing more than nine ounces each. There is another peculiarity about the tarn which deserves notice. The north-east portion, from Miss Postlethwaite's meadow to the boat-house, an area of two or three acres, is covered with a dense mass of coarse grass, intermixed with flags, &c., their tough matted roots forming a covering or sward, the substratum composed entirely of a semi-liquid mass, which if the sward were removed, could not be traversed even by a dog or cat. But for this mass of felted roots it would be impossible to cross it at any time, and although there is not much danger of breaking through the covering of tangled roots, timid people have no business upon it.

It may be objected that Clerk's Beck has been the inlet of Urswick tarn for hundreds, perhaps thousands of years, which is no doubt true, but when we consider that before mining commenced in Furness, scarcely anything but clear water entered the tarn by this source, very little solid matter could be deposited, and it is probable that as much sediment is thrown down in one year at the present time as was formerly in a hundred years.

Near the south-west corner of the tarn, in cleaning out a deep drain, a Roman tripodal vessel of bronze, in perfect condition, holding about a quart, was discovered, and is now in the possession of the Rev. T. Tolming of Coniston. After passing the church we come to Kirkflat Tarn, in a circular basin about two acres in extent, of no great depth, and which about twenty years ago was for a few weeks quite dried up, when several interesting things were discovered, amongst them teeth of three different species of deer. Little Urswick Tarn, formerly abounding with medical leeches, is three or four hundred yards beyond, but we fear the creatures have been annihilated. There is a good quarry of limestone close to the tarn, where a few good specimens of coral have been obtained, but Organic Remains are not abundant. A few years since in excavating stone for lime burning, they discovered in a cleft of the rock, several pieces of ancient weapons, as swords, spears, &c., which were sent to a museum at a distance. We now enter Little Urswick, the head quarters of the parish, almost surrounding the village green, on the high side of which is situated the ancient Free School, our first and only school, where the Rev. William Ashburner, vicar of Urswick presided. Our parson was a kind and good master, and at the annual "barring out," quietly submitted to have a little dirty water thrown over him when he attempted to storm our barricade, and as he never could succeed, being always forced to retreat, it was not with frowns and threatenings for another time, but with a good-natured smile at his defeat. We were then free to choose our captains and clerks for the annual foot-ball match, who were also our commanders for the cock-fights on Shrove-Tuesday, at the ancient cock-pit ring, on the upper side of the green in the centre of the village. In the *holy* and *humane* pastime carried on at this ring, on Shrove-Tuesday, of the years 1796—98, the vicar presided in gown and bands, and might be seen running round the ring, being then a strong and active man, but instead of preaching a sermon, or reciting passages of Scripture to his unruly parishioners to keep them quiet, he used a more persuasive and irresistible argument,— a nice little two-handed cudgel, as we have seen.

We are not to presume from the above sketch, that the rev. vicar was a bad man, on the contrary, from our own knowledge he was a kind-hearted Christian minister, a scholar and a gentleman, beloved by his pupils and generally by his parishioners. But the times are altered, seventy years have made many changes, and although cock-fighting is now abolished by Act of Parliament, a portion of the cock-pit ring still remains on Urswick Green; perhaps Jonathan Oldbuck would have described it as an interesting Roman antiquity, surrounded by a ditch and vallum.

On the low side of Urswick Green, and directly opposite the school may be seen the very humble cottage described in a former part of this work, as containing a small geological collection of Organic Remains, made by a child seventy years ago, strange to say, several of them were not only of unknown species then, but are unknown and undescribed even now. These were principally collected from the material excavated from a well behind the school where a small quantity of the *debris* is still lying. At this place, when a child, we have spent many days turning over and breaking up pieces of clay and soil, seeking for and carefully taking home beautiful Encrinital remains, &c., to enrich our little store. Some of the same species of minute fossils may be found there yet, by carefully turning over with a pick or spade, portions of the rubbish, and although there was a considerable quantity at first, it is now nearly exhausted, therefore we will hasten on to Hawkfield, which we believe to be the head quarters for these beautiful forms in England, and where we shall be kindly welcomed by W. Cranke, Esq.,* the owner, who, although not an active geologist himself, takes a lively interest in our pursuits, and kindly permits us to dig in any part of his land, and he will particularly direct your attention to the south end of a long, narrow, pasture field called the "Bank," where he has sunk a well, the material excavated from which has been strewed around the place. This *debris* has yielded thousands of beautiful Organic Remains, but a few may still be found at any time, particularly after heavy rain. We will now suppose ourselves at the place intending to work in earnest. Our first operation is to examine the clayey ground around the well, but we need not spend much time in this, as Mr. Cranke will

* Now deceased.

cheerfully lend us pick, spade, or clay-tool, to dig here and there in the field at some little distance from it, and this should be done in the following manner:—Mark out a space about three feet by two, cut the sods and carefully put them on one side, take off the soil which will be ten or twelve inches in thickness and place it aside also. Immediately under this is a tough blue clay without Organic Remains, extending in some parts of the field to a considerable depth, and in others not more than sixteen or eighteen inches in thickness, having beneath it, a white fossiliferous clay containing thousands of beautiful fossils and of a great many different species, chiefly very small, some so minute as to require a lens for perfect examination. This white clay is not continuous over the whole field, but occurs in patches here and there, over a space of about half an acre, at the south end, and in excavating, if we do not come to the white clay at about two feet below the surface, we may conclude there is none at this place. We now fill the hole up, and carefully replacing the sods, make another similar hole three or four yards from the first, where it is likely we shall be successful, if not, try again, and we shall be sure to succeed in the end. A basket will be necessary, and when we have found the white fossiliferous clay, we must cut out with our clay-tool, moderately large pieces, and take a basket-full home unbroken; after which to find those beautiful forms, proceed in the following manner. Break *quietly* the large pieces into others about half the size of a hen's egg, take out the larger Encrinital remains, which will principally be stems and sculptured plates of the head, put the small pieces into a washing-pot, then take them to the pump, or still better to a running stream, fill the pot with water, stir gently the pieces of clay, till the water becomes as muddy as possible, pour off *quietly*, then fill the pot again, repeating the operation until the clay is entirely washed away, and the water comes off perfectly clear, The residuum will consist of particles of sand, gravel, and thousands of beautiful minute Organic Remains, which should be spread out on plates to dry, and the larger fossils taken out, then with a common lens carefully examine the remainder, and you will be delighted with the result of your labours. Many of these Organic forms are as small as the point of a needle, yet even they, on close examination, will be found to have holes through them.

We have been thus particular because any stranger going to Hawkfield without this information would think it a very poor place, and after seeking on the surface of the ground around the well in the Bank Field, might only find a few inferior specimens and go away disappointed, but if he follow the instructions given above he may obtain an interesting collection of these beautiful forms. The fossils procured from this deposit of white clay possess one peculiarity,—all their sculpturing are as sharp and perfect as when they had life, and motion—from which circumstance it is reasonable to infer, that they have not been transported from a distance, but are at, or near, the place where they became living forms. It is still more remarkable and mysterious when it is considered that in the same field there are lying on the surface, several travelled blocks of Mountain Limestone scattered here and there, forming in one part of their range a curious wave of stone which seems to have been suddenly arrested in its progress, and in some degree resembles the high-water mark of the tide on the sea-shore. All the transported blocks are composed almost entirely of Encrinital remains, many of the stems or columns being nearly as much in diameter as an ordinary spade shaft, but all of them as geological specimens are utterly worthless, their markings being faint and obscure, and although these blocks may be considered almost a solid mass of Encrinital columns, yet we never could find a single plate of the head. Besides, the limestone rock may be seen *in situ* within thirty yards of the place where we have obtained the greater part of our private collection.

This extraordinary circumstance induced one of the first Palæontologists of the age, after having seen those beautiful sculptured plates, which are portions of the head of the Encrinite, to recommend that we should seek diligently in the rock nearest to where they were found, and we should to a certainty discover perfect heads of the Encrinites, desiring to be informed the first, if such were the case. We followed his advice, and engaged a quarryman to break up two or three of these travelled blocks of Encrinites, but instead of perfect heads, we never found the smallest portion of one. The gentleman alluded to was at that time connected with the Government Museum of Practical Geology, and certainly he had reason to think that perfect heads would be found when he recommended us to search, but he was mistaken, and the mystery remains even greater than before. In addition to the place described above as favourable for obtaining these beautiful Organic Remains, there is a small private quarry at Hawkfield deserving a visit, which is not in constant work, but occasionally as stone may be required for fence walls, and we believe produces a greater variety of Organic Remains than any other quarry in Furness; amongst these may be included thirteen or fourteen species of *producta*, four or five of *spirifer*, with *orthis*, *orthoceratite*, *belerophon*, *cornulite*, &c., and it is the only place in Furness where we have found the *perma*. We have also obtained a new species of *spirifer*, striated with very close and beautiful lines, perfectly parallel, and finely waved longitudinally, about the size of a pigeon's egg. If this quarry were extended on the surface by baring more of the rock, many good specimens would be developed, and probably new species might be obtained. Before leaving Hawkfield we may just examine the buildings of the ancient farm of Bolton, which gives name to one of the townships of the parish of Urswick, and is now attached to Hawkfield, also occupied by M. Cranke, Esq. Those buildings contain the remains of the ancient chapel or chantry, of Bolton, granted by Robert de Denton, Abbot of Furness, to Sir Richard, son of Sir Alan de Coupland, and his heirs. The chapel buildings are indicated by Gothic arches over doors, windows, and various ecclesiastical details, but they are now converted to other purposes, as sheltering cattle, &c. Leaving Hawkfield and Bolton, we proceed by Beckside on the road to Scales, but before we reach the village a lane branches off to the right, leading to Mere Tarn, at the south end of which is a soft, " swaggy" patch, similar to that at Great Urswick Tarn, where it is said, a man who went to shoot wild fowl, sank through the soft spongy sward, and was never seen again. About five or six hundred yards south of Mere Tarn are the extensive ruins of Gleaston Castle, the residence of the Lords of Aldingham after their original home in Aldingham was swept away by the sea, the exact place of which cannot now be determined, but it certainly now forms a part of Morecambe Bay. A short distance before we come to the castle yard, a small quarry may be seen on the east side of the road, distant about one hundred yards, belonging to the estate of Gleaston Castle, the property of His Grace the Duke of Devonshire, and occupied by Mr. Ormandy as tenant. This quarry has not been in work for many years, and all the *debris* has been cleared away so as to expose a comparatively level floor covered with a bed of carboniferous shale eighteen or twenty inches in thickness, abounding with thousands upon thousands of beautiful and well preserved fossil shells, several with interior muscular markings, some with both valves in contact and otherwise perfect, showing when ground through the outer shell and polished, the form of the fish entire in the inside, and particularly the spiral process of a small bi-valve, hitherto classed as an *orthis*, but from the fact of its having a spiral process must be removed into the genus *spirifer*. We have now arrived at this little quarry, and before we begin work we may be permitted to make a short digression, as it is somewhat connected with the place. In September, 1851, after spending a morning geologising at the great limestone quarries of Skipton, in Yorkshire, we proceeded direct to Paris, and in twenty-four hours found ourselves geologising in the Gypsum Quarries of Montmartre, the scene of Baron Cuvier's extraordinary discoveries, where we devoted two most unsuccessful days—not finding anything worth taking up. We afterwards visited the National Museum of France, in the Jardin des Plantes, and spent several hours in the geological department of that establishment, making a minute examination of nearly all the Organic Remains contained in the place, and in a glass case set apart for superior and scarce fossils, we were delighted to find a few of our old friends of the same species as those so abundant in the shale bed at Gleaston Castle, but they were all inferior specimens, small, and mutilated, and such as we would not have taken up from the ground. We happened to have a few rather good specimens of the same species, from Gleaston Castle, in our pockets, as well as a few of the beautiful Encrinital Remains from Hawkfield, after calling the attention of the curator of the museum to this particular case, and comparing them we presented the whole of ours to him, for which we received a thousand thanks, and he exultingly placed them in the same case near the others, but apart from them. He was most struck, however, with a very perfect specimen of *spirifer squamosa*, and was extremely anxious for us to point out on a large map of England, the exact place where they were obtained. We mention this circumstance as an instance amongst several others, that we possess beautiful

I

natural productions in our little kingdom of Furness, which although either overlooked or slighted by the natives, are highly appreciated by strangers.

It will be necessary to procure pick and spade for digging and turning up the shale, and these implements may be borrowed at the farm buildings below. When a portion has been cut out by the spade, it will divide into flakes, and you will be pleased to see an abundance of well preserved fossil shells of many different species developed between every flake, the lines and other sculpturings as sharp and clear as when they were living animals. You may now with three or four hours' careful working, obtain a moderately good collection of Organic Remains, as the carboniferous shale at Gleaston Castle contains, with few exceptions, nearly all the different species peculiar to the Mountain Limestone, to be found in Furness. It will be desirable, besides collecting the detached and loose fossils, to gather a few of the best flakes studded with fossils in high relief; these flakes, after the "quarry damp" is dried up, will be persistent and very good specimens for the cabinet. Before going home we must not neglect to visit the great and pure spring of Gleaston Castle, undoubtedly one of the greatest in the north of England. It issues from the high land on the east side of the farm buildings, and is sufficiently copious to turn a mill, even from the spring head, without storing in a dam for that purpose. We do not know a spring that can be compared with it, unless it is perhaps the "Anchorite Well" at Kendal, "the river Jordan" of the Baptists, wherein that denomination of Christian worshippers immerses its members, which is supposed to be the best spring in Westmorland, and may be considered the nearest to it, yet after all is very inferior.

ULVERSTON TO LINDAL AND STAINTON.

For this ramble it will be advisable to go by the morning train from Ulverston to Lindal, where you will arrive in a few minutes. A short distance from the Railway Station is Lindal Cote, the site of the Ulverston Mining Co.'s iron ore works, especially interesting, for, besides being a good locality for obtaining Mountain Limestone fossils, it is also the place where in sinking a pit or shaft for the water-level from Lindal Cote to Urswick Tarn, at the depth of 40 feet from the surface they cut through a light drab coloured material which floated on water when thoroughly dry, containing many pieces of unfossilized wood and other vegetable remains, consisting of seeds, seed vessels, &c. At the time of its occurrence, we wrote a short paper on the subject, which after keeping in our possession for seven years,— from 1855 to 1862,— we presented to the Geological Society of London, where it was introduced by the President of the Society, and read at their meeting, May 24th, 1862, in the following abridged form:—

"This deposit * has been sunk through during the progress of works undertaken by the Lindale Cote Iron Ore Company for drainage purposes. The mines are situated in the well-known hæmatite district of Low Furness, about three miles S.W. of Ulverston, in a valley between two ranges of low hills belonging to the Mountain Limestone series. The physical geology is varied in character, — a fine sequence of the following beds in descending order from the Upper Silurian occurring in the hills lying north of this valley, viz., Lower Ludlow Rocks, Upper Ireleth Slates, Lower Ireleth Slates, Coniston Grit, Coniston

* On a Deposit with Insects, Leaves, &c., near Ulverston, by John Bolton, Esq.

Flags, Coniston Limestone (equivalent to the Bala Slates), and Green Slates with Porphyry, which last rocks extend northward for many miles beyond the boundary of Furness. South of the valley in which these mines are situate, the Mountain Limestone is developed on a large scale, being upwards of six miles in breadth. The exact position of Lindal Cote Mine, upon the promontory of Furness, is about halfway between Morecambe Bay and the estuary of the Duddon.

"In sinking shafts to enable them to excavate a waterway from Lindal Cote to Urswick Tarn, in 1855, down the course of a valley lying about 100 feet below the table-land, and receiving the drainage of about 600 acres, a deposit of greenish-drab clay, six feet in thickness, was met with at a depth of forty feet from the surface, in the shaft nearest but one to the mines, and at the highest 'level.' This clay-bed contained pieces of unfossilized wood, associated with numerous leaves, seed-vessels, and other vegetable remains. Among the few which can be determined are leaves of Beech, with the epicarp of the fruit receptacle, and a well preserved branch of Sphagnum. A few well preserved insects also occurred in the deposit. Of these some have been determined by Mr. Stainton, F.G.S., as fragments belonging apparently to a land Hemipterous insect, and one as a portion of an Orthopterous wing. Three nearly perfect specimens of Apterous Hemiptera he referred to *Cimex*, or an allied genus. Microscopical examination of this clay shows us the conditions under which it was deposited.* It is seen to be chiefly composed of lacustrine *Diatomaceæ*, the facies of which point directly to a mountain-tarn as the origin and support of their existences. The list of forms obtained from it is nearly paralleled by those which Dr. Balfour and other gatherers

* To Miss E. Hodgson, of Ulverston, is due the credit of examining this deposit for *Diatomaceæ*, and mounting the specimens that are here referred to.

Fig 1.
Section of a Shaft at Lindale-Cote Mines near Ulverston.

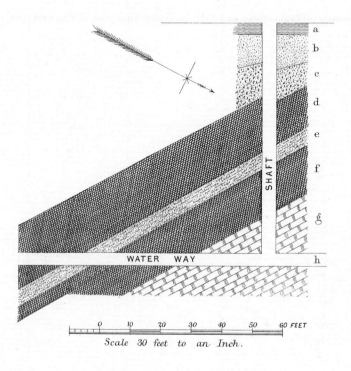

Scale 30 feet to an Inch.

a. Soil, 3 feet.

b. Pinnel (Rubble) 10 feet.

c. Gravel, 12 feet.

d. Black muck, 16 feet

e. Clay bed with vegetable matter and Insect-remains, 6 feet.

f. Black muck, 12 feet.

g. Limestone, 12 feet.

h. Water way.

of *Diatomaceæ* have obtained from subfossil clay and peat deposits in the Mull of Cantire and elsewhere. The genera represented ar *Geomphonema, Tribunella, Epithemia, Surirella, Cocconeis, Cyclotella, Pleurosigma, Campylodiscus, Navicula, Tetracyclus, Odontidium, Cymatopleura, Cymbella, Stauroneis, Pinnularia, Synedra,* and *Eunotia.* These have been kindly determined by Dr. Wallich, F.G.S. Siliceous spicules of freshwater sponges also occur in this deposit.

"The length of the water-way driven from the mines to the tarn is a mile and a quarter; and in the portion tunnelled twelve vertical shafts were sunk at convenient distances — nine in the bare Mountain Limestone at the lower end of the adit, and the remaining three through the overlying 'drift,' which at No. 10 shaft was thirty feet in thickness, at No. 11 sixty feet, and at No. 12 thirty feet. It is therefore evident that a basin in the limestone was crossed by the line of the work.

"Probably these clays have a considerable extension to the N.E. and S.W.; for thin beds of the same deposits were met with in a trial shaft sunk by the Lindal Cote Company, at the highest part of the table-land, one-third of a mile S.W. of No. 11 shaft, and at the same level. Here, as in the first-proved locality, the clays yielded vegetable remains and *Diatomaceæ.*

"The eight feet of gravel alluded to in this section is of the ordinary alluvial character, made up of water-rolled pebbles of Upper and Lower Silurian rocks, bedded in quartz sand. There is elsewhere evidence of this drift-deposit having resulted from north-westerly currents.

"From the lowest part of the soft limestone thus pierced, a horizontal 'drift' was driven northward in search of iron ore, and in progress of the work, it was found that the limestone and the lowest superimposed beds had a steep downward inclination, also, that the plant-bearing deposit, when cut through by the gallery, had thickened to fifteen feet. The wood imbedded in the lower seams of the clay was partly converted into a soft, blue pigment, having phosphate of iron for its colouring-matter.

"Thus it appears evident that the areas anciently covered by the lake-water were those of the long valleys which course sinuously between the low hills of Furness.

"A second adit, driven southward from the bottom of the shaft, cut into a good bed of iron ore at twenty feet from the commencement.

"Glancing backwards for a moment at this scant record of a local and comparatively insignificant deposit, I diffidently claim a value for it in any scheme cast for the determination of Pleistocene time. In the absence of great and sudden cataclysmal irruptions of water which could fill valleys with drifted material, and of which I conceive we have no settled evidence, it appears to me that the time required for the deposition of this great thickness of nearly 100 feet of transported material upon the comparatively flat surface of this lacustrine clay by the ordinary degradation of the low hills around it must be one far extended beyond our ordinary notions. The material of which the whole thickness of the superimposed deposit is composed is of strictly local origin, and, in the absence of violent sweeps of north-lying water, and sudden fillings-up, by such means, of the shallow valleys by the locally derived detritus, I am at a loss to see how the distribution could have been effected, except by ordinary aqueous and pluvial agencies extended through a long period of time.

"P.S.— Since the above paper was communicated, the miners have exhausted the iron ore in the pit, section fig. 2; and then they sank to a further depth of about thirty feet, but without getting through the soft limestone. They have now left it altogether, and have sunk another shaft about 220 yards to the north of it; and at about the same relative depth they have found the same deposit, contain-

Fig. 2.
Section of a Shaft at Lindale Cote Mines near Ulverston

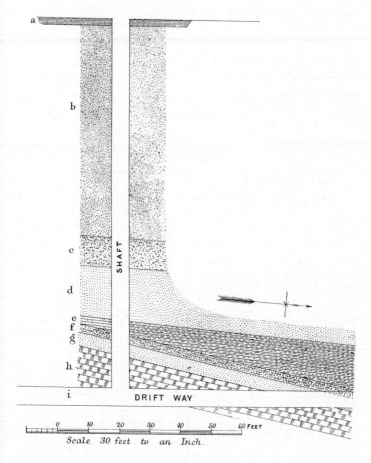

Scale 30 feet to an Inch.

a. Surface Soil and Roadway, 1 foot.
b. Hard reddish rubble (Pinnel) 68 feet.
c. Gravel 8 feet.
d. Yellowish sandy material, 16 feet.
e. Greenish Clay with plant remains, 3 feet.
f. Clay coloured blue in patches by phosphate of Iron, and with woody fragments, similarly coloured, 2 feet.
g. Sand 6 feet.
h. Very soft sandy Limestone, abounding with characteristic Mountain Limestone fossils, 22 feet.
i. North Drift, with Iron-ore in the Limestone.

ing vegetable remains, &c., but not in abundance. The miners say also that they found the same material in a shaft about 200 yards north from this new shaft, that is, about 420 yards north of No. 2 section. If this be correct (and I have no reason to doubt it), it demonstrates that the deposit covers a triangular area, the three sides of which are respectively 420, 450, and 600 yards in length. — May 24, 1862, J. B."

A long and animated discussion followed, in which several members took part, however, no theory was propounded to account for its origin, and it was agreed to ask the author's permission that the specimens sent up to London, should be forwarded to Professor Hier, at Zurich, as being the most competent person in Europe to deal with the subject. They were sent accordingly with a copy of the paper, and Professor Hier, after a most minute and careful examination, wrote a very able article, and figured several of the specimens, publishing them subsequently in a Zurich periodical in the German language, a copy of which is now in the Geological Society's Library in Somerset House. This strange woody deposit is now proved to extend over an area of at least 200 acres, having been met with in sinking nearly all the iron ore pits at Lindal Cote. It is always associated with a white deposit of decomposed Mountain Limestone, abounding with characteristic Organic Remains, most of them very perfect, but soft, crumbling, and almost worthless for the cabinet. We have obtained good specimens of the genus *Bellerophon*, at the railway tunnel, near Lindal Cote, and one moderately good specimen of the *Eumphales crustatus*. At Stainton, one mile further, there is a good locality for Organic Remains. About a quarter of a mile before we come to the village, we pass a field on the left, called "Cowstone," where G. Huddleston, Esq., sank two pits, cutting through in the course of the work a thick deposit of alum shale, containing a great variety of Organic Remains, the prevailing type of which was of the genus *Pleurotmaria*, &c. The most perfect specimens of these fossils were not persistent, for being highly charged with alum, they soon became covered with an efflorescence of that crystalised material, and in the course of two or three years fell to pieces, therefore were not fit for the cabinet. At Stainton, however we have a limestone quarry where fossils, though not abundant, are by no means scarce, all showing interior organisation, and susceptibility of the highest polish. Blocks of exquisite marble, containing three or four different species of Organic Remains, may be obtained sufficiently large for ornamental table tops, especially beautiful when the whole is one mass of coral surrounded by a border of the natural rock. This is one of the best, if not the very best Mountain Limestone quarry in Britain, occupied by Mr. Garden, of Dalton, of whom mantel-pieces may be obtained, not manufactured by himself, but supplied by him to the marble manufacturers in blocks of stone, cut to the required dimensions, and squared up in the rough, which are sent to Liverpool, Manchester, Preston, Lancaster, and Glasgow, where they are cut and polished by machinery, and made into beautiful mantel-pieces. We have purchased three of these from Mr. Garden, one of which we have put up in our own house, where it may be seen by any one who takes an interest in the natural productions of Furness. The marbles yielded by the Stainton quarries are known in distant places as "Ulverston marble," which has almost entirely superseded that of Dent and the quarries of "New England," near Burton-in-Kendal. The two latter were once in high repute, but the Ulverston marble has now thrown the produce of both out of the market. Perhaps the nearest approach to the Ulverston marble in excellence is the Purbeck marble, a good example of which is seen in the fine columns of

Temple Church, London, but even these, although they have the same beautiful dappled grey colour, are very inferior to ours, having a partially "pitted and punctured" ground, from which that of Stainton is entirely free.

We must again make a short digression to show that we neglect and almost despise elegant natural productions of our own and procure from a distance, at a great expense, an article in repute, when we have a superior one of the same kind almost for nothing at home. We will give an instance in proof. We were requested by a clergyman from the neighbourhood of Shrewsbury (who claimed to be a descendant of Sir George Graindorge), to visit Furness Abbey and minutely examine the sepulchral slab of his ancestor, now lying amongst the ruins of the abbey, and if possible to say from whence it came. We went to the abbey accordingly, and took from the tombstone, now broken in two pieces, a small fragment which we ground up and polished, and from the Organic Remains it contained, we demonstrated that it was Purbeck marble, and had been brought some hundreds of miles from its parent quarry. However we did not neglect to let the gentleman know that we had tens of thousands of tons which could be obtained at a trifling expense within two miles of the abbey, and of a quality superior to that of his ancestor's monument. Perhaps the heirs and executors of Sir George had followed the example of the abbot and monks of St. Mary's, for we have ascertained to a certainty that the small marble columns in the vestibule of the Chapter House of the abbey are Purbeck marble also, which has been procured at a considerable outlay, and brought from a great distance, whereas if the abbot had given orders to his vassals after matins, he might have had a sufficient quantity of marble from his own demesne at Stainton, for all purposes, and of a superior quality, laid at the abbey gates, before the bell for vespers echoed its note through the lovely "Vale of Nightshade." No doubt the abbot and monks of St. Mary's were men of great learning (as were all the Catholic clergy of Europe), but as the infant science of geology had no existence at the time of the foundation of the abbey, and as the examination of rocks was no part of the duty of these holy men, we may charitably suppose they were entirely ignorant of the fact that tens of thousands of tons of beautiful marble were deposited within two miles of their gates. It may be objected that this was "a long time ago,"—that the world is much wiser now,—that many wonderful discoveries have been made since then, and a case like the above could not occur at the present time. We will give another similar instance, which is comparatively recent. The late T. R. G. Braddyll, Esq., when he rebuilt Conishead Priory, brought from the south of England many ship-loads of Bath Stone, at a very great expense, when he had a very superior building material in his own park close at hand, without carriage, which could be laid at the works in quantity sufficient to build a hundred mansions at a trifling expense. We believe the estate in the immediate neighbourhood contains sufficient of this splendid building material to erect a city as large as London and Paris put together, and we presume the rejection of his own beautiful building stone in favour of an inferior and very expensive one from Bath, was not only a great mistake, but a positive misfortune, and may be considered one of the causes of the decline and fall of an ancient and universally respected Furness family, the Braddylls of Conishead Priory. To return to Stainton. About 200 yards west from Stainton Head, and abutting on the south side of the road from Stainton to Dalton, is a small meadow nearly level, except the end next the highway, for an area of about a quarter of an acre, which rises rather suddenly, and forms a small hill, composed entirely of quartz sand, which is dug out for building purposes, and a considerable quantity has been taken away, so that one side of it forms a sand cliff, eight or ten feet in height. About seven years ago a person, excavating in this breast of sand, laid bare one side of an ancient British cinerary urn, which he thought would be full of gold, and was so anxious to be at it that he had not patience to remove the sand from about it, so as to enable him to take the urn out whole, but seizing it roughly pulled it to pieces, (it was composed of half-baked clay mixed with something like pounded brick) but instead of being a "pot of gold," it was filled with partially calcined human bones, and also contained a small bronze instrument something like a spear head. A communication was immediately made to the Superintendent of Police, who sent an officer to take possession of them and bring the whole to the Police Station, believing they would discover a murder or something of the sort, but the Magistrates of Ulverston were not ignorant of cinerary urns, and cheerfully gave them up to us, and we not only obtained all the pieces of the urn the police had collected, but we afterwards secured most of the fragments which had got into other hands, and with great labour and care put them together again. The urn was about the shape and size of a common straw bee-hive, and was set with the mouth upwards, covered with a rough undressed flag of Upper Ireleth Slate. About two years after this event, another urn was discovered in the same way, at the same place, holding human bones, with a smaller urn inside, about as large as a pint measure, also filled with bones. With some difficulty we managed to obtain and dress these up like the first, and they are now in our possession. The large ones have a raised border round the mouth, five inches in breadth, ornamented in a very rude manner with diagonal lines from left to right, crossed with others from right to left, the design of both urns being nearly the same. This dry sand hill has been considered a suitable place for interment, and we have no doubt more urns will be discovered before all the sand is taken away.

From the quarry at Stainton may be obtained several specimens of the genus *cornulite*, particularly *Turbinola fungites*, all susceptible of a most brilliant polish, many entirely detached from the matrix in which they have been imbedded, and as they are free from crystalisation, the polishing displays their interior organisation in a very beautiful manner, some appearing as if the different flakes of which they are composed could be separated like those of a salmon when dressed for the table. Polished geological specimens are both interesting and instructive objects for the cabinet and beautiful ornaments for the mantel-shelf, and as the operation of polishing may be considered a pleasant amusement rather than otherwise, and may be conducted comfortably by the kitchen fire-side without annoyance to anyone, we may be permitted to give specific instructions for the process. In the first place, procure a piece of gritty flag stone about half a yard square, this, with sand and water, will be proper for grinding down the specimen to be polished, until it is level and true, although the face will be left entirely covered with rough scratches. Another piece of flag with a finer grain will also be required, which should be used like the first, as it will take away all the deep scratches, and leave the face of the specimen covered with others much finer. Now take Water of Ayr stone, which is perfectly free from grit, use it in the same manner by rubbing backwards and forwards until the scratches have been taken out entirely, be very careful to have pure water only, and that no sand touches the specimen, and in a little while it will have a fine, smooth, level face, every way suitable for polishing. Water of Ayr stone may be obtained at any of the marble works in a large town. These operations may be considered outdoor work, as no particle of sand should be near the polishing place, for a single grain would make a mark on the specimen. The last operation is polishing, for which purpose it will be necessary to procure a smooth piece of deal board, about eighteen inches in length and twelve in breadth, a piece of new, soft, coarse, woollen cloth, rather longer than the board, tacked at each end to the under side, so that it may be entirely covered with cloth, and an ounce or two of oxide of tin from the chemist. We will now suppose the operator to be comfortably seated at a small table by the fire-side, with a cup half full of clean water, a teaspoon, and a small scoop made from a quill, for putting on pinches of the oxide of tin. Being now ready, first pour a teaspoon full of water on the cloth-covered board, and a quill full of oxide of tin, then commence rubbing backwards and forwards, or round and round as may be most convenient, until the tin is exhausted. Renew the tin and water five or six times as described, wash and wipe the specimen to see what progress has been made, when the smooth face of the stone will be found to have begun to brighten, which is the first indication of a true polish. Continue the operation on the cloth, rubbing as before, and replacing the water and tin as often as necessary, and with one hour's patient working a brilliant and lasting polish will have been effected, and if it has been kept free from sand and other foreign matter, will come out a splendid specimen polished like a mirror, without a spot or scratch to detract from its beauty, making an ornament for the cabinet of any gentleman in England. All the corals and cornulites from Stainton, when polished, show interior organisation, and offer a highly instructive lesson to the student.

ULVERSTON, LINDAL MOOR, MARTON, DALTON, TO ST. HELEN'S.

The first part of this ramble may be considered theoretically as a traverse between the boundary of the Mountain Limestone and the Lower Ludlow Rock. We will suppose the student to commence his journey from the centre of the town, proceeding up the Gill Banks, taking the course of the Town Beck, by the old powder magazine which was built and used as the receptacle for the powder and other ammunition stores of the first Ulverston volunteers, raised in 1803, at the time of the threatened invasion by Napoleon I. They consisted of four companies of strong, active young men, numbering upwards of 300 rank and file, splendidly equipped and clothed, having amongst other expensive articles superfine scarlet coats, the finest naval blue cloth pantaloons, fitting tight to the leg, and braided with red cord down the outside seams. The battalion was instructed in military exercises and evolutions by retired veterans from the regular army, but we will say no more about them at present, as we may speak of them again in a future article. Immediately above the foot-bridge in the Gill Banks, there may be seen in the beck bottom a small patch of Mountain Limestone, and a little higher up the stream an out-crop of Old Red Sandstone. There is here evidence of extensive denudation, as it is probable that these two formations, at some former period of the earth's history, extended over a considerable area, for we find two similar patches half a mile from this point, in the bottom of the small beck at Rosshead, flowing down the next valley to the west, and two miles further, in the Powka Beck stream, at the point where the "tail-race" from the old iron works at Orgrave flows again into the stream from which it is diverted at a higher level. This outcrop of Old Red Sandstone at Powka Beck, is overlaid by the usual old red conglomerate, but not in any great force. We believe there is no place in Furness except those above mentioned where the Old Red can be seen and examined *in situ*. The junction between the Mountain Limestone and the Lower Ludlow Rock (Coniston Flag), is in no part of their range a straight line, but toothed or notched into each other in a curious manner. Three or four hundred yards west from the Gill Banks we find the Mountain Limestone again at Tarn Close, where there is now a limestone quarry. The produce of this quarry is a very good building stone, fine grained, close, sound, and durable. Organic Remains are scarce, but we have obtained a rather good specimen of vegetable fossil at this place, being a portion of a large *calamite*, five inches in diameter. From Tarn Close, proceed westward about a quarter of a mile to Rosshead, and in the beck bottom by the road side may be seen the patch of limestone alluded to above, and in the bank on the other side of the road Old Red Sandstone. Pass these as not being worth your notice, as you are now close upon the best place, not only in Furness, but we believe in Europe, for one particular class of Upper Silurian fossil, *Cardiola interrupta*, *Pterinea*, &c. The identical spot where they were found is the middle of the public road, or street of the hamlet. The road at this point is composed entirely of the soft mud stone, which Mr. Salter calls "Wenlock Shale," the correctness of which name, and those of some other of the rock formations of this district are now disputed, so that it is desirable, and even necessary, that a new classification of the different stratifications of Furness be taken so as to set that matter at rest. However, let the name of the Rosshead formation be what it may, it is a deposit of high geological interest, and has yielded numerous splendid fossils of the Upper

Silurian system, therefore, we may be permitted to enter into some detail respecting it. M. Barrande, the eminent Bohemian geologist, presented a few of his best specimens of the *Cardiola interrupta*, from the neighbourhood of the city of Prague, to the Museum of Practical Geology in Jermyn Street. We also gave two or three good specimens to the same institution, and although ours were by no means the best we had obtained from Rosshead, they were very superior to those from Prague. This strange fossiliferous deposit of mud stone covers but a small area, not more than about twenty-five yards in length by ten in breadth, and is now almost exhausted. It had a saddle or anticlinal appearance, rising quickly on one side, and descending quickly on the other. Silurian fossils were first discovered here by the "road men," in taking off the crown of the little hill to repair the road at another place, who having no knowledge of geology, destroyed many beautiful fossils in their ignorance, remarking " these queer things were very plentiful, one or more turning up almost with every stroke of the pick, but they carted them all away for repairing the road." As this small hill required to be excavated deeper to make the road perfect, we cheerfully undertook it for them at our our own expense, having obtained permission from the surveyor of the parish roads to do the work in our own time, and in our own way, provided we did not stop the traffic. For this purpose it was necessary that only one half of the breadth of the road should be excavated at once. However, we set about it forthwith, determined to excavate the whole place with hammer and chisel, one foot deep, and split up every piece of rock, carefully examining it before we threw the fragments on one side, so that we might find everything it contained. It is acknowledged, and has been noted and commented on, by more than one of our fashionable novelists, that "the disciples of Isaac Walton" possess a considerable amount of patience, and the true "Waltonian" has been given as a type or living personification of the Goddess of Patience herself. The above remark, no doubt, is true in part, and the angler with his rod and creel may spend a few hours now and then by the river side, but the angler's boasted patience ranks very low indeed, when compared with that of the fossil hunter, who will not only spend hours and days, but weeks and months at a single quarry, and almost at the same rock. Many examples might be given in proof of this superior patience and perseverance, but one will be sufficient, as it is so ridiculously extravagant, that it is almost beyond belief, yet is in every part literally true. Let the reader picture to himself an old man over seventy years of age, an inveterate fossil-hunter, quietly folding up his old wallet, made from coarse canvas wrappering, for a cushion, and sitting down on it in the middle of the street, or town-gate of Rosshead, subject to the jeers and witticisms of every passer by, and with hammer and chisel patiently dig and split up the soft rock of which the road is composed, and continue his work from morning till night without a moment's rest, and without meat or drink during the whole day, not even once rising from the ground to straighten himself. When night comes see him pick up his wallet, hammer, and chisel, and trudge home, having excavated a portion of rock about four feet square and one foot deep, without finding a single fossil worth taking home. He will be surprised however to see him again the next morning, seated on the road as before, working with the same patience and perseverance to return at night with hopes unabated, but without a single fossil. The third day passes with the same result, the old man's patience is not yet exhausted, and he will be seen again on the fourth morning at the same place, and at the usual time. The result of the fourth

K

day's labour is somewhat better than the other three, the old man succeeds in finding two or three rather indifferent specimens of *Cardiola interrupta*, but none of them good. It might naturally be supposed the man's endurance was now fairly ended, and he would give up the place as a failure for Organic Remains, — such however is not the case, for he continues to work one, two, or three days almost in every week, for upwards of two years, and during that time spends more than two hundred days, from morning till night, sitting in the middle of the road working as described above. The reader will by this time perceive that the old man alluded to was the writer of this sketch. After we had taken down the whole crown of the hill about one foot deep, and carted away all the *debris*, we recommenced to excavate and take off another "flake" or lift, one foot thick, just as we had done at the first. This also with great perseverance we accomplished, but in it Organic Remains were very scarce, yet we would not give it up until all was finished, and although it had the appearance of being a most discouraging and wearisome task, it was not without its pleasures, being rewarded now and again by turning up a beautiful *Cardiola interrupta*, or a good specimen of some other species of fossil, and we were often amused by the remarks and speculations of the lookers on, who became very anxious to know what we were seeking so industriously, (as we took care none should ever see us find a fossil,) some said we were seeking for indications of iron ore, some for lead, others that we were seeking for gold, but one shrewd miner said "nay, nay, I dar say ye are o rang, for't aald fella hes summat off at we kna nout about;" probably none of these people ever heard of a fossil in their lives. On another occasion we were very much amused with the conversation of some school-boys. It was one bleak dull morning in January, with a sharp cutting wind from the north-east, and snow had fallen during the night, about three inches deep, but the snow was no obstruction, as we only wanted one square yard for a day's work, so scraping it off with our shoe we sat down to our work in the usual way. This was earlier than common, we had been at work some time, when the boys came by snow-balling each other as they went to Pennington School. The morning being rather colder than usual, we had to work hard to keep ourselves warm, and we succeeded very well, for, besides finding two or three beautiful *Cardiolas*, we discovered two new species of Organic Remains. It was getting dusk, and we were preparing to go home, when the school-boys from Pennington came by on their return, and one of them suddenly exclaimed " sees'ta Tom! by gock, if t'aad fella isn't sitten ther yet, an lile Seppy Atkisson hes brout his mudders chair cushon out and is sitten beside em ta keep em cumpeny, an ise varra near starv't to deeath wi cummin yam, an t'aad fella was sitten here when we went ta't school in't mornin." Little Seppy Atkinson, our companion, lived close by where we were working. He was a sweet little boy, often sitting down beside us for an hour or two, amusing us by his innocent prattle, and on this occasion, when we arose from the ground to go home, we could not stand for a little while, but had to be supported for a few minutes by the little boy. Now when it is considered that this extraordinary fossil hunting at Rosshead, extended through two winters, it must be acknowledged that we displayed more than an ordinary amount of patience, some may call it foolhardiness, which is indeed partly true, yet when we look back to the time, it affords us pleasure even now; however, we do not consider it any great achievement, or work to effect any really useful object, but on the contrary, as a great waste of valuable time which might have been spent to better purpose. The student must not infer from the above remark that we disapprove of fossil hunting altogether, for it is not the case, but there is a fascination in the pursuit which concentrates the mind to that particular object, and has a tendency to draw to extremes, as in the above instance. If seeking for Organic Remains, be kept within reasonable bounds, we know of no earthly pursuit equal to it, and for this purpose we have for many years carried a small hammer and chisel in our pocket, as a gentleman carries a snuff box, or a diamond ring on his finger. We think these tools would be much more useful, and no degradation to him, and we are not ashamed to say those articles have many times travelled with us to Ulverston Church, for which beautiful fabric, and for our beloved pastor, the late Rev. Canon Gwillym, we had the greatest veneration.

From Rosshead to Lindal Moor there is nothing of geological interest; our traverse is in a direction west-south-west, roughly indicating the line of junction between the Mountain Limestone and the clay-slate. The boundary between these formations cannot be ascertained the whole distance, yet we have the true junction well defined at both ends of the traverse, viz. :—Rosshead and Lindal Moor. The immense deposit of hæmatite ore at Lindal is, without doubt, the most extensive in the kingdom ; millions of tons have been taken away, and many millions yet remain, indeed it may be considered the centre or head-quarters of the mining district of Furness. The mines here give a great variety of facts and specimens bearing on the question of the formation of hæmatite ore, and should be visited by all geologists, especially by those who may be inclined to grapple with that mysterious subject. It is a remarkable fact that none of our eminent geologists have hitherto attempted to solve this problem. Mr. Binney, a high geological authority, read to the Literary and Philosophical Society of Manchester, a valuable paper on the AGE of the hæmatite ore of Ulverston and Cleator. The paper contains much interesting matter, and proves that the author is well qualified to deal with the question, and also, that he is intimately acquainted with the hæmatite district, and it is especially valuable as establishing the comparative AGE of the hæmatite, with reference to the other formations with which it is associated, but neither Mr. Binney, nor Professor Philips who read two papers to the British Association at Leeds, on the hæmatite ore of Ulverston and Cleator, have touched on the question of the FORMATION or ORIGIN of the hæmatite, or any other variety of the ores of iron, or what are the conditions presumed to be favourable for their formation. On the north-west side of Lindal Moor, on the brow of the hill, may be seen a number of hollows or depressions ranging in a straight line parallel with the range of the clay-slate, and at no great distance from it. They have the appearance of being trial works for ore, but they are in reality natural "sinks," or "swallows," and indicate the line of a cavern in the limestone, or between the limestone and the clay-slate. The water from some of the mines is raised by power to the surface, and conducted into one of these natural "swallows" and is seen no more. In a driftway in one of the pits they have recently broken into another cavern, which also ranges north-west and south-east and ends at a natural "swallow" similar to the others, in the bottom of a deep basin near Marton. It is somewhat remarkable that, although the water from several of the mines at Lindal Moor has for generations been disposed of in this way, none of the receptacles was ever surcharged, neither is it known, if the water ever issues to the surface again. A short distance south-east from Marton, and abutting on the mineral tramway, there is a small quarry in the Mountain Limestone, which has yielded several splendid specimens of vegetable fossils, all being characteristic coal plants, viz., *Sigularia, Lepidendron,*

Calamites, &c. This quarry also yielded moderately good specimens of Organic Remains characteristic of the Mountain Limestone, *i.e.*, *Spirifer*, *Producta*, *Orthis*, *Corals*, *&c.*, and we have also obtained one good specimen of *Mitchliena grandis*. Vegetable fossils have not been found at any other limestone quarry in Furness, except a single specimen at Tarn Close. It is a strange circumstance to find *Spirifer* and *Orthis* on the same slab of limestone, and almost in actual contact with *Sigularia* of the coal measures. At Marton we again come to the junction of the Mountain Limestone and the clay-slate, the line of boundary between these two formations passing through the centre of the village. About two hundred yards to the north of Marton, by the road side, there crops out a soft, earthy rock, used for repairing the roads, containing two or three different species of *Graptolites*, all in high relief on the surface of the stone. This is the only place, with one exception, where we have ever seen *Graptolites* having a palpable substance, those from all other places seem as if they were painted on the stone. There are here also other specimens of Silurian fossils, but no very good specimens can be obtained. From Marton our traverse is south-west to Orgrave, where we have a small patch of Old Red Sandstone in the brook (Powka Beck), and at a point about two hundred yards down the stream an underground communication commeneces it is said with Yarl Well, near Dalton. The entrance to this passage is somewhat above the ordinary level of the water in Powka Beck, therefore, the stream in the cavern is not constant, but much more copious in floods than at other times. A short distance south from this point are the extensive iron ore works of J. Rawlinson, Esq., at Cross-gates. Some parts of the mining operations at Low Cross-gates consist of open works, *i.e.*, an immense hole or quarry open to the day. In the cliff forming the western

IMBEDDED LAYER OF PEAT.

side of the deep excavation, and from twenty to thirty feet below the surface, may be seen a soft, white deposit like flour; this is a mass of perfectly decomposed Mountain Limestone, containing upwards of a hundred tons in weight, in the middle of which is imbedded large pieces of decomposed wood, one of which is twenty inches in diameter, and must have been part of a large forest tree. Small pieces of the wood are converted into a beautiful blue pigment of phosphate of iron. At High Cross-gates, six or seven hundred yards north from this point, there is another large excavation, which has been wrought as an open works mine, in the cliff forming the north end of this immense chasm, and fifteen feet below the surface, there is exposed a bed of black peat moss eighteen inches in thickness, containing vegetable remains of several different species, with the *Elitra*, and wing covers of three or four species of insect, and the horn of a large species of deer, the base or lower end measuring nine inches in circumference. The latter, with some other interesting matters from this deposit, were kindly presented to us by Mr. Rawlinson at the time of their discovery. To account for the overlaying of this peat moss to a depth of fifteen feet with a foreign substance, would have been no very difficult matter, supposing High Cross-gates to have been the lowest point of drainage of a water-shed of three or four thousand acres, but seeing there has been no violent cataclysm, or wave of translation to cover it up suddenly, and as it has been deposited quietly, and laid conformably to the strata beneath, as well as the covering above, and, although High Cross-gates is the lowest point of drainage, yet from the physical conformation of the adjoining land, it must have been very small indeed, and evidently could never have exceeded an area of more than ten or twelve acres in extent. Therefore, the occurrence of a mass of peat in this extraordinary situation is very perplexing, and may be considered a very difficult problem to dispose of by geological reasoning.

Half way between High Cross-gates and Low Cross-gates, three hundred yards from each, but not in a direct line, Mr. Brogden has commenced some new iron ore works, or rather extended the present works at Ure Pits to this point. In a "bore hole," No. 24, they have proved a section of the earth's crust so extraordinary, that we must not omit to give it in detail. It is as under:—

	Feet.	Inches.
Surface	1	6
Pinnel	24	6
Decomposed Limestone	17	0
Yellow Clay, mixed with Iron Ore	4	0
Black Mould	4	0
Iron Ore	2	0
Black Mould, mixed with Iron Ore	6	0
Iron Ore	8	0
Decomposed Limestone	7	0
Black Woody Deposit	12	0
Decomposed Limestone	6	0
Black Mould and Wood	2	0
Yellow Clay, mixed with Ore	11	0
Ditto mixed with Iron Ore	6	0
Black Mould, mixed with Iron Ore	10	0
Black Mould	4	0
Black Mould, mixed with Ore and Limestone	3	0
Total	128	0

About two hundred yards west of Low Cross-gates are the extensive iron ore works of the Barrow Hæmatite Iron and Steel and Mining Company, at Mouzell. Portions of the works have been wrought open, and a few years since, when all the drift and other worthless material were cleared off, and the iron ore rock laid bare, the surface afforded a geological lesson of more than ordinary interest. The greater part of the face of the iron rock presented the appearance of having been subjected to heat, that had not

Section of a Bore-hole at High Cross Gates.

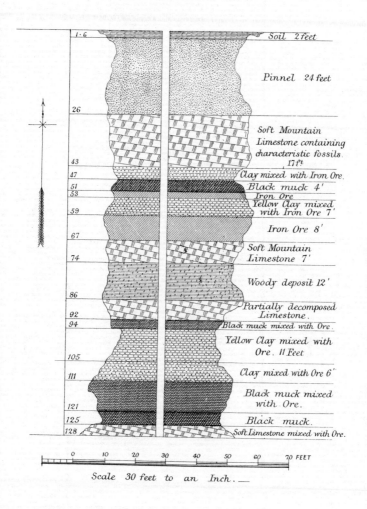

Scale 30 feet to an Inch.

been general over the whole mass, but quite local, proceeding from several centres, and its effects seemed clearly traceable for a yard or two on the continuous rock, when it died out entirely. Adjoining the south-west end of these works are the iron ore mines of Geo. Ashburner, Esq., of Elliscales. In one of the pits, at a depth of 112 feet, they have found many pieces of unfossilised wood with other indications of the same woody deposit as that at Lindal Cote and Cross-gates, thus proving that it has a range extending over an area of several hundred acres. On reviewing all that has been stated concerning this woody material, and especially that most mysterious section of the earth's crust, proved by a "bore hole" at Mr. Brogden's new mines between High and Low Cross-gates, we must consider that to reach the first or highest woody deposit we pass through two beds of Mountain Limestone, one 17 feet in thickness, the other seven, and to reach the lower accumulation we pass through three, the aggregate thickness of which is 30 feet. If we offer to account for the deposition of this strange material so deep in the earth, the mind is lost in wonder, and we are utterly helpless. Geology offers no assistance in the matter, and the most accomplished geologist may wander in the dark from one theory to another, and with all his learning and research, return to the point whence he set out, unable to throw a gleam of light upon the subject, for certainly it involves a geological difficulty of no common kind and is shrouded by an impenetrable mystery. Let us then, instead of presumptuously attempting to interpret this problem by forcing and twisting facts to make them harmonise with some theory of our own, humbly acknowledge our entire ignorance of it, believing it to be one of the secrets of nature not yet revealed to man. When shafts are sunk at or near to "bore hole" No. 24, Cross-gates, some interesting geological facts will be brought to light. Mr. Brogden is a lover of geology, (as all iron ore masters should be, for the successful carrying out of mining operations depending on geological reasoning), and besides having made considerable progress in the science, liberally encourages geological pursuits in others, therefore, we have no doubt he will give to the geological world the benefit of his discoveries.

Doodles Quarry, situated on the north side of Dalton, is an instructive place, for, although Organic Remains are neither good nor plentiful, there is an object of interest which should be noticed. It is not in work at present, but there is a double limekiln in good repair, and in the walls forming the outer crust of the kilns may be seen several specimens of perfect black and perfect white limestone united as if they had been two liquid streams of carboniferous matter flowing into each other, and so thoroughly combining that they cannot be separated even with the hammer. These would make beautiful ornaments when ground and polished. There is also a small quarry at Elliscales, where some of the Organic Remains are weathered and decomposed in a singular manner, particularly *Turbinolia fungites*, and other species of the genus *Cornulite*. Many of the specimens are skeletonised, and may be compared to a house that has all its interior wasted and destroyed, except the framing timbers and rafters, in others again, the whole interior is gone, leaving nothing but the outer wall or shell singularly perfect through its whole length, showing every line and flexure with perfect clearness. These latter the poetic, or imaginative, may call "fairy cups," with as much propriety as the worthless fragments of Encrinital columns of Holy Island are called "Saint Cuthbert's Beads," by Sir Walter Scott, in his poem of Marmion.

On the south side of the town of Dalton, there issues from under the high cliff near the church yard, a copious spring of water flowing at the rate of upwards of a thousand gallons of very pure water per minute. Near to which may be seen in the face of the rock, a broad "dyke" of carboniferous spar, bearing north and south; in its northward range it will pass under the Market Place of the town, but we have not made any excavation in that direction sufficiently deep to meet with it again. At the point where the spar commences near the spring, the stratification or bedding of the rock is almost perfectly horizontal, but at the tunnel of the railway, within one hundred yards of this place, it has a considerable inclination, and dips to the south-east at an angle of about 35 degrees. At Crooklands, near the east end of Dalton, are the new iron ore mines of Messrs. Clegg and Co., and Messrs. Denney and Co., the latter having yielded a few specimens of Organic Remains, all perfect hæmatite, and of the same specific gravity as their best samples of that material. Mr. James Denney has obtained a specimen, at present in our own possession, which is a portion of an Encrinital column with a beautiful sculptured surface, all its markings sharp and clear, and it may be used as an element in reasoning on the question of the formation of the hæmatite iron ore.

Half a mile north-west from Dalton is the interesting quarry of Saint Helens, particularly noticed by the late Mr. Jopling, in his "Sketch of Furness and Cartmel." We will give his list of Organic Remains found at this place without attempting to extend it, yet, we have no doubt a careful investigation would enable us to do so. However, we may be permitted to correct, in the kindest manner, a trifling error with reference to his *Astrea Annana*, which is without doubt the *Nematophylum arachnoidum*, and *N. minus*, (Mc.Coy). Many beautiful specimens of this interesting fossil may by obtained here without much trouble, some weathered in a curious manner, but all showing interior organisation, and a susceptibility of the most brilliant and durable polish, every way suitable, rough or polished, for a geological cabinet, neither would they disgrace any gentleman's mantel shelf, but, on the contrary, be elegant and instructive ornaments calculated to create a desire for beautiful natural objects.

ULVERSTON, KNOTT-HOLLOW, IRON YEATS, GROFFACRAG SCARS, GAWTHWAITE, TO KIRKBY.

From Ulverston proceed one and a half miles north-west to Knott-hollow, where a small quarry in the clay-slate formation has been in existence for generations, and as it has now become a place of great interest, we offer no apology for giving some particulars respecting it. The quarry is situated high up the eastern slope of the hill, 782 feet above Ordnance high water mark, and 107 above Knott-hollow Tarn, an oval sheet of clear water about eight acres in extent, lying immediately below it. From the site of the quarry we have a splendid and extensive view, on the east the town of Ulverston in the foreground, beyond, Morecambe Bay, and in the distance, Ingleboro, and the mountain range which constitutes the "back-bone of England," and southward, the rich agricultural district of Low Furness, with the Isle of Man and some of the mountains of North Wales in the distance, when the weather is favourable. The "dip" of the rock is to the south-east, and the angle of the dip much greater, than the inclination of the hill itself, being about 80 degrees with the plane of the horizon. Knott-hollow formed part of Osmotherley Common, (recently enclosed), and the quarry is set apart for all the inhabitants in the township of Osmotherley, in the parish of Ulverston, and although it is free to all to get stone for their own use, without interfering with each other, yet it is not in constant work, but sometimes deserted and undisturbed for years.

Our experience at this interesting quarry will afford an instructive and encouraging lesson for the young geologist, leading him to persevere and not despair of finding Organic Remains, where they have never been seen before, as in the instance of the Skiddaw Slate, which, hitherto considered almost entirely unfossiliferous, now produces another class of Organic Remains, some new and others never previously discovered in Britain. We had in the course of a few years made several visits to Knott-hollow Quarry, but were unsuccessful in finding the least trace of Organic Remains, yet we were not much disappointed, as we invariably enjoyed with the greatest relish, the beautiful and extensive panorama by which we were surrounded. The mountains and the sea — always most attractive objects — almost commanded our worship, and in the middle distance in the plain below, the scene of our childhood, the farm house in which we were born (Mountbarrow) was a point of especial interest, and sometimes, in a dreamy and forgetful mood, we would endeavour to make out the form of our father in the fields near the house, although he had been dead upwards of sixty years. Let no one say that if he go out for a particular purpose and be somewhat disappointed, there can be no enjoyment in anything else. At the time of our first geological visit to Knott-hollow, the quarry had laid silent and neglected for many years, but in 1857, the owner of the spade forge at Rosshead, constructed a weir for a reservoir, a little below Knott-hollow Tarn, for which we had to take a series of levels. At this time he had obtained and placed near the quarry, about one hundred tons of stone, all in flaggy pieces, something like drain covers, not in a heap, but spread on the ground in cart loads, precisely as we had long wanted. We therefore began at one end of this great mass of stones, and, examining carefully every one on the ground on both sides, as well as on the edges, succeeded in finding two specimens of a Silurian Encrinite, *Actinocrinus pulcher*, the fossil in both entirely decomposed and reduced to an impalpable powder of a brown colour resembling snuff, while the place in which they were imbedded was quite sound, and when washed at a running stream the fossil was dispersed entirely, leaving the matrix as a perfect mould, from which we took a gutta percha cast in high relief. We were delighted to see all the markings and sculpturings as clear and sharp as could be executed on copper or steel by the engraver, and the mould not injured in the smallest degree, so that hundreds of casts may be taken if required. The lines of sculpturing on all of them are beyond anything for beauty we have ever seen in stone of any kind. Some of the markings are so small as to require a lens to examine them properly. Soon after, Osmotherley Common was enclosed, stone was required for fence walls, so the quarry was worked again for a considerable time, and we did not neglect to visit it two or three times every week and carefully examine nearly every stone used in the construction of all the fence walls required for the enclosure. In this manner we obtained several good specimens of the *Actinocrinus*, some containing two or three almost perfect heads of that beautiful and scarce fossil. We also found in the centre of one of the nodular concretions, a partially decomposed *Cardiola interrupta* — rather softened, but the most perfect specimen we had ever seen. We then began to scrutinise with more care the whole of the Kirkby Moor range, from Knott-hollow northwards by Iron Yeats, Groffa-cragg Scars, and Gawthwaite, to the great quarries at Kirkby, belonging to the Duke of Devonshire. We now propose to take this range, and examine all places of interest returning to Knott-hollow to finish our ramble. Proceeding northwards, one mile, we come to Iron Yeats, where a quarry in the clay-slate exists, yielding principally coarse flags, lintels, and walling stones, and although there are considerable excavations, the place is not always in work, but like Knott-hollow, sometimes unused for a year or two. We need not expect much here, as we have visited it several times and never found anything worth taking away, although we have met with *Actinocrinus* and one or two corals, but all decomposed and worthless, both fossils and matrices. The dip of the rock at this quarry is at a lower angle than at Knott-hollow, and varies in different parts of the quarry, in one place it is about 60°, in another not more than 45° with the plane of the horizon. At Groffacragg Scars, one mile further, there is a small excavation made by two labouring men who get a few flags occasionally. The stratification of the rock here dips to the south-east at an angle of about 80°, and we meet with our old friend *Actinocrinus*, but specimens are very scarce, and we have not found a single good one, most of them being small fragments of the stem or column. We have also discovered one or two fragments of *Cardiola interrupta*, and a rather good specimen of a Silurian coral, but nothing of much interest. From Groffacragg Scars northwards, one mile, are the quarries at Gawthwaite, rather extensive in their excavations, and always in work. In one of our visits to this place, accompanied by a geological friend, in a remote and unfrequented part of the quarry, we came upon a mass of soft, decomposed rock, at least half a ton in weight, presenting fossiliferous indications, and on breaking it up it displayed one face almost entirely covered with the heads and stems of *Actinocrinus*, several being almost perfect. This at first sight seemed a prize, but it soon proved to be almost valueless, as it was so soft that it would break to pieces with the slightest blow of the hammer, and out of the whole mass we did not obtain one good specimen. However, as we have found *Actinocrinus, Cardiola interrupta*, &c., with two or three different species of coral at all the places enumerated above, which extend nearly the whole length of the Kirkby Moor range, we may assume it to be all one formation, although we call one division Lower Ludlow Rock, another Upper Ireleth Slate, and a third Coniston Flag, yet, without doubt, they are all the same, and for reasons before given we feel compelled to class the whole as Coniston Flag. If we search carefully all the extensive quarries of Gawthwaite, we may perhaps find a specimen or two of *Actinocrinus*, but probably they will be decomposed and worthless, as the quarries have been in existence for hundreds of years, and many thousand tons of *debris* are lying on the ground. If these immense heaps be examined carefully, we shall be certain to find a few specimens of coral of two or three different species, all of which we believe are new, and generally the centres or cores of the nodular concretions existing in considerable numbers amongst the refuse, some fifteen or twenty pounds in weight, when split open exhibit now and then one with a coral in the centre. These corals are all very hard, durable, and susceptible of a high polish, some being very beautiful. After leaving Gawthwaite, we proceed south-west round the northern point of the Kirkby Moor range to the immense slate quarries belonging to His Grace the Duke of Devonshire, at Kirkby, ranging along the western slope of Kirkby Moor in a south-south-west and north-north-east direction for upwards of a mile. A visit to these interesting works is highly instructive, and much may be learned, we will not, however, trust to our own description, but give an extract from "Jopling's Furness and Cartmel."

KIRKBY SLATE QUARRIES.

"Far off may be heard the deep sound of the blast, as it echoes among the mountains; and upon a nearer approach, the rattling of the rubble and stones, as they tumble and leap down over the sides of the hills, the incessant clatter of the slate-rivers' hammers,—'tinkling animation, noisy concussion, and thundering explosions' —give to these manufactories of slate a great appearance of activity. They form, too, a very prominent object, and their heaps of rubbish are a sombre appendage to the hills in which they are situated. The romantic positions in which they occur will repay a visit, and more especially so if the tourist be geologically inclined; for there is a rich and extensive field laid open for investigation, by the immense excavations showing the positions of the strata, joints, lamina of depositions, and the planes of cleavage.

"In noticing the structure of the slate, one of the most singular features appears to be the cleavage, which maintains nearly one unvarying bearing and angle with the horizon, however the stratification may be contorted. The principal vertical joints have a north and south direction; and, in conjunction with the beds or planes of stratification, and the cleavage planes, divide the body of the slate into rhomboidal masses. Some of the north and south joints continue to a great distance, and are champions, or master joints, crossing everything. These are frequently filled with a variety of minerals, containing in a matrix, generally of carbonate of lime, small quantities of yellow copper, lead, iron pyrites, black jack, manganese, and sulphate of barytes; slickenside also accompanies them. In some of the strata, on the plane of deposition, are rows of nodular concretions, formed around a nucleus, the nature of which has not yet been determined; but they bear, in the best specimens, a resemblance to the *Cephalaspis Lewisii* of Agassiz.

"The quarries are worked on the side of a hill, and

mostly open at the top, though some are subterraneous, resembling mines; levels, or tunnels, being formed to bring away the rock and rubbish, and to allow the water to drain off.* The slate is detached from the rock by means of blasting, in which operation the quantity of powder used on different occasions varies considerably. All the arrangements of boring and charging having been completed, the signal for the workmen to retire is given by the vociferation of the word 'fi-er,' which is immediately answered by all tools being laid down, and retreat being made to a place of security. When not accustomed to the occurrence, the feeling of suspense, between the lighting of the fuse and the explosion, is rather painful, and the nervous action considerable; especially when it is known that a large quantity of powder is employed — a quantity indeed, which, upon some particular occasion, may amount to a barrel. At last the fuse hisses — bang goes the blast — the rocks vibrate — crash! crash! crash! come the huge masses down, tumbling over each other, as if in play: pieces, some tons in weight perhaps, are flung against the sides of the quarry; other pieces take an upward flight, and, after an aërial voyage of a few hundred feet, plough up the ground, and continue bounding onward until their force is broken. It rarely happens, however, that such powerful blasts as these occur, the general charge being only a pound or two of powder; but when the rock is worked into a particular position, one such will bring down an immense quantity.

"The operation of boring and blasting each large piece thrown down has then to be gone through; after which the whole is reduced, by means of sledge-hammers and wedges, to such pieces as may be conveniently carried away. The next process is that of riving, or splitting the

* The largest level in this district has recently been completed at the 'Burlington Slate Works,' at Kirkby.

rock into thin plates. This part of the manufacture is carried on at the outside of the quarry, and requires much dexterity in the workmen; the art is, indeed, only to be acquired by an apprenticeship. The plates are then formed or dressed into slates, and classed according to their size and thickness. The old classification by the workmen was, London's, Country's, Tom's, Peggy's, and Jerry-tom's; London's being the best, and the others constituting the gradations."

At the southern extremity of the long range of quarries belonging to the Duke, is another large one occupied by Messrs. Postlethwaite Brothers, Slate Merchants, Kirkby, which has yielded several specimens of Organic Remains, particularly our old friends, *Actinocrinus, Cardiola interrupta, &c.* None were good specimens, being soft and decomposed, and the markings were indistinct, although yet sufficient for establishing the identity of the species, and demonstrating that these Organic Remains are common to all the principle quarries on Kirkby Moor, and characteristic fossils of the Coniston Flag. Our route is now southward, a mile or two, to examine two or three small trial quarries on the western brow of the moor, facing Soutergate, where we have obtained several good specimens of three or four different species of *graptolite*. Thence eastward to Harlock, where a small quarry used for obtaining road material, yields stone all decomposed and fragmentary, containing several species of Organic Remains, including those we have before enumerated, and in addition, fragments of *Obtusa caudata, &c.*, but soft and worthless. We have now made a circuit of nearly the whole of Kirkby Moor, and examined all its principle quarries, proving that almost all the Organic Remains found at any one quarry are common to the others, and those at Knott-hollow very superior to any. Returning to Knott-hollow, there is a second quarry, which we purposely omitted to mention until we had seen what Organic Remains may be found at those just described. This quarry, also free to the inhabitants of the township of Osmotherley, is very small, being not more than half an acre in extent, and situated about 200 yards west from the other. At the first Knott-hollow quarry, the stone requires blasting, and a regular quarryman must be employed for that purpose, but in the small one the rock is looser, and may be got with ordinary quarry tools without powder. At this place we have obtained by far the best and largest specimen of *Actinocrinus* we have ever seen, therefore we may be permitted to say a few words about it. One day having visited the larger quarry to find it deserted, and all the stone carted away, we proceeded westward to the smaller one, where two men were at work, and we observed indications of Organic Remains. Borrowing an iron bar from the men, we began work for ourselves, and soon discovered a prize, but it required great labour and patience to develop it carefully, as the plane of deposit is diagonal to the cleavage of the rock, therefore, large pieces had to be detached, as the fossil only covered one end of the slab. We carried home two pieces in our arms, not trusting them either in wallet or basket, for fear of injuring the beautiful sculpturing of the fossil. The next day, with the help of our son, a boy of fifteen, we excavated other portions of the same rock, which we carried in our arms to cut and reduce at home, where we had proper tools for the purpose. We wrought day by day, for eleven successive days, until we had excavated the whole mass of rock covered with *Actinocrinus*, and, although we took considerable care in our work, if we had known that none of the fossil remained behind, we should have taken even greater precautions to prevent scratches, or injury of any kind happening to this splendid group of Organic Remains, comprising a great number of nearly perfect heads and

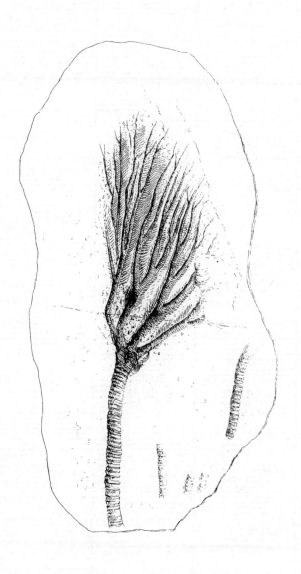

ACTINOCRINUS PULCHER
From Knott-hollow Quarry,
near Ulverston.

stems, or columns, of *Actinocrinus*. We have with considerable labour cut out and reduced it to five inches in thickness, yet, it is still between two and three hundred pounds in weight, and although it is inclosed in a gilded frame, 42 inches by 28 inches, it is by no means a cabinet specimen, or suitable for a gentleman's private collection. This splendid fossil has been inquired after, and probably its final destination will be the British Museum. If casts in gutta percha be required from any specimen of *Actinocrinus* of Knott-hollow, it is a great advantage that all the animal matter be entirely dissolved and gone, leaving the matrix as a perfect mould, as it enables us to take a cast in a few minutes at one operation, which will be in high relief, showing the finest lines of sculpturing with the utmost clearness, whereas, if the fossil had been in relief it would be necessary to take a cast from the original fossil, and then a second from the first, being much less perfect than if taken at one operation.

GEOLOGICAL WANDERINGS IN CARTMEL.

It was intended originally to confine our geological wanderings and remarks to High and Low Furness, but as a considerable part of this work has been in manuscript for several years, during which important changes and improvements have taken place on our eastern boundary, it has been suggested that we ought to introduce a short sketch of the parish of Cartmel, especially the southern part of it, Kents Bank, and that interesting and beautiful watering place the lovely village of Grange.

Grange, from its exceedingly picturesque and sheltered

situation, having an aspect to the rising sun and being perfectly sheltered from the north and western storms, has a decided advantage over all the watering places of the west coast of England, and besides being a delightful summer residence, is also a snug retreat for the winter months. Its rocky sea-shore promenade has no rivals in the kingdom, and for at least 300 miles has no superior, except Allonby, in Cumberland, and as a geological field it is superior either to Allonby or Saint Bees.

We will now attempt to mark out one or two pleasant geological rambles of easy accomplishment, in which ladies may join without much fatigue or discomfort.

We will suppose the party to set out from the splendidly appointed hotel, taking the direction of the sea shore southward. After passing through the village we come to a comparatively smooth floor of Mountain Limestone rock which is almost continuous to the southernmost point of the land at Humphrey Head, two miles. Organic Remains, for the first mile, are rather scarce and difficult to develop. We pass Kents Bank and Abbots Hall, the beautiful seat of J. S. Young, Esq., the owner of Kirkhead Cavern and the whole range of Kirkhead Hill. The cavern is a most interesting place and well deserves a visit. About ten years ago we formed a small party, consisting of Messrs. W. Salmon, F.G.S., John O. Middleton, and four others, to explore and excavate in the floor of the cavern. We wrote a short paper on the subject a few years ago, which was presented and read to the Anthropological Society of London, in the following form:*—"Although no remote antiquity can be claimed for this inhabited cavern as a dwelling-spot, yet the record of its contents cannot fail to

---

* "Interesting discoveries in the Kirkhead Cave, near Ulverston." At a meeting of the Anthropological Society of London, a paper was read communicated by Mr. John Bolton and Mr. G. E. Roberts, on the Kirkhead bone cave, near Ulverston, and numerous specimens of the bones discovered there were exhibited.

be of anthropological value. The geological history of the cavern is simple. Caverns in limestone rocks belonging to the carboniferous series are numerous, wherever that formation is developed; in most cases they have communications with the surface, either by a fissure or cleft in the strata, or in stratigraphy of the rock. The cave is situated on the western flank of Kirkhead Hill, on the west shore of Morecambe Bay, at a point about six miles from Ulverston. The hill rises abruptly from the sea-shore, within a quarter of a mile of high-water mark, to a height two hundred and sixty-four feet, and is composed of Mountain Limestone. The entrance to the cavern is eighty-five feet above high-water mark, the inclination of the hill from the cavern mouth downwards being 65°. Beyond the mouth, the height of the roof varied from eighteen feet, at the part nearest the entrance to twelve feet, the length of the cave was found to be forty feet, and its width twenty feet, the area consisting of one irregularly oviform chamber. No communication between the roof and the surface of the rock above was apparent, though the thickness of the brushwood which clothed the hill rendered any investigations difficult. From the shape of the cave it appeared to have been a natural reservoir for waters permeating through the rock, both from the surface and from springs, such communications having been extinct long before its occupancy by man and the smaller carnivorous mammalia. The floor of the cavern, when thus first visited, was composed of a brownish red indurated clay. Two labourers made excavations in this to the depth of eight feet over an area of about fifty square feet. The clay contained many angular fragments of Mountain Limestone, probably fallen from the roof of the cavern, and a few pebbles of Upper Ireleth Slate, or of Coniston Flag, varying in size from a walnut to an orange, and derived probably from rocks which are situated northward.

These were all water-rolled. There was also found in the clay a considerable number of mammalian and bird bones. At the depth of four feet a portion of the right parietal bone of a human skull was thrown out. The bone was sound and strong, being evidently part of the cranium of a full grown human subject, and a little lower the radius and ulna of the right arm of a young subject not more than ten or twelve years of age. Continuing the excavation to a depth of eight feet, another human bone was obtained which proved to be the second lumber vertebra of a full grown human subject. Below the cave-earth is a floor of stalagmite. Very recently further diggings have been made into the cave-earth; amongst the articles obtained at various depths on this occasion were several jaws of badgers, and other bones of that animal, together with bones of fox, wild cat, goat, kid, pig, and boar; and, at a depth of three feet a large and strong humerus of man. Mr. J. P. Morris, who accompanied the author, found three human teeth and fragments of human bones, together with a tusk of wild boar, and a portion of a large deer horn, a foot in length, and ten inches in circumference at its extremity for articulation with the skull. Scattered through the clay were many fragments of stick burnt at one end, as if from the remains of fires; these, though interspersed through the whole mass were more abundant toward the bottom of the deposit. In the stalagmite were several pieces of wood-charcoal firmly imbedded in the mass with carbonate of lime. A Roman coin was also found a few inches below the surface of the floor of the cavern."

Near the south end of Kirkhead Hill, in the deep rock-cutting made by the Ulverston and Lancaster Railway, there is exposed a strong band of black indurated carboniferous shale dipping to the south-east in accordance with the rock above and below, but at a somewhat higher angle. Proceeding south along the shore of Morecambe Bay,

Organic Remains are more plentiful, and large blocks of limestone are strewed on the loose shingle, many composed almost entirely of coral of several different species, all susceptible of a brilliant and lasting polish, and numerous specimens may be obtained without much trouble. Before we come to Humphrey Head Point, black shale crops out on the shore containing numerous fossil shells of many different species, all characteristic of the Mountain Limestone, as *Orthis*, *Spirifer*, the most prevailing types being *Cornulites* and *Productas*, but no very good specimens of either are obtainable. We may now turn round the extreme southern point of Humphrey Head, on the rock if the tide permit, as there is no danger because we have a "safe side" *i. e.*, if we cannot go forward we can retreat, which is not always so when any one is unfortunately surrounded by the tide.* Immediately after we have rounded

* A lamentable case of this kind on these sands recently happened. Four youths, brothers, the eldest not twenty years old, sons of a poor widow in Staffordshire, her entire support, but out of work, had travelled on foot for want of money, and had been to Barrow to seek amployment, on their return they attempted to cross the sands at a wrong state of the tide, and sad to relate they were surrounded. When the water reached them they did not give up all hope of saving themselves, the eldest brother supporting and encouraging the youngest but one, the second brother doing the same by the youngest, and thus they struggled on for some time, but the youngest, although he had been partially held up by his brother, sank and was carried away, and soon after his supporter sank also. The eldest kept to his precious charge a while longer, at length he was forced to let go, and in the last extremity was saved in the following manner. The perilous situation of these youths having been observed from the western shore of the Ulverston Sands, a boat was at once pushed off and entered by two daring young men, an exciting race commencing between this small boat's crew and the ruthless tide. Every instant of time was of the last importance, as it involved the chances of life or death for these poor young strangers, and was literally and truly a race for life, but when the boat arrived at the scene of disaster, three of them had been swept away and perished, and the only survivor was rescued just as he was sinking, and lifted into the boat as life was nearly extinct. Those three poor strangers now sleep side by side in a remote corner of Trinity Church-yard, Ulverston. Their funeral was attended by a crowd of people, none of whom were either friend or relation except the poor widow and her only son, who with difficulty supported her feeble steps to the grave to take a last farewell of her darlings. The scene had a saddening effect on all, and many of the spectators departed from the grave with tearful eyes.

the southern point of Humphrey Head, we see an inscription recording the death of a young gentleman who was killed by a fall from the summit of the rock, and a little further, but close to the foot of this immense limestone cliff we come to the "Holy Well of Cartmel." This is a small spring of mineral water issuing from the base of the rock, said to be very efficacious in gouty, rheumatic, bilious, and cutaneous diseases. From a recent analysis by Mr. E. T. Thorpe, of Owen's College, Manchester, the water has a temperature of 52·7 F. Sp. g. 1005·13, and contains

Chloride of Sodium	331·7524
Sulphate of Calcium	88·4898
Chloride of Magnesium	43·4882
Sulphate of Sodium	24·3971
Carbonate of Calcium	9·2029
Sulphate of Potassium	9·1749
Silice acid	1·2341

and traces of phosphates, bromides, iodides, fluorides, and chlorides of Ammonium, Calcium, Magnesium, Lithium, and Sodium.*

Proceeding westward on the sea-shore, about a quarter of a mile, we find a strange deposit of Permian Brixia, at Hougholm Point. The north end of this immense mass rises perpendicularly eighteen or twenty feet from the surface of a level meadow, and dipping south-south-east at an angle of eighteen degrees, continues to slope in that direction until lost under the sands of the bay. This Brixiated deposit is composed principally of angular fragments of all the different rocks of the district including fossils and hæmatite iron ore. From Hougholm westward to Coupren Point, and from thence northward by Cark to Quarry Flat, there is no rock on the sea-shore, nor anything of much geological interest, and it is not yet determined whether the rocky framework below the surface is Mountain Lime-

* For a full account of its virtues and history, see an interesting article by Dr. Barber, in *Baines' Lancashire*, new edition, Gent, Manchester.

stone or the Lower New Red Sandstone, probably it is the former, and if so the sandstone of Holker Park has no great range southward, otherwise there would be a likelihood of finding coal in that direction, especially as a thin band of that mineral crops out near Quarry Flat. At the latter place the sandstone cliff abounds with vegetable fossils, all characteristic coal plants, viz., *Sigilaria, Lepidodendron, Calamites, &c.*, many of the specimens containing their lanceolate leaves *in situ*, bristling round the stem of the tree as if growing.

From Quarry Flat Point northwards, a most interesting natural section extends along the western side of Holker Park, developing four distinct changes in the rock escarpment as under. At Quarry Flat Point, Sandstone abounding with vegetable fossils (coal plants), northward about eighty yards a strong bed of Coal Shale with two bands of *Iron Stone* each about two inches in thickness, and a fossil tree imbedded in the cliff. One hundred yards beyond this, a strong band of Mountain Limestone crops out containing numerous characteristic fossils *(Encrinites, Orthis, &c.,)* the range of limestone continuing about one hundred yards, when it suddenly becomes sandy, and in three or four yards more, changes entirely into perfect sandstone again, abounding with fossils of a different character altogether, such as marine fossil shells of several different species, and numerous strong vermicular markings, but not one vegetable fossil in the whole range of the sandstone cliff. This second range of sandstone cliff extends a considerable distance, when the Mountain Limestone crops out again and continues to form a fringe on the shore of the bay, uninterupted for two or three miles. To entirely develop the fossil tree alluded to above, it would be necessary to take away all the superincumbent rock and other material, when it would be seen lying nearly in a horizontal position on the ground, and be a very interesting object, but it would effect no really useful purpose, otherwise we have no doubt the Duke of Devonshire would order it to be done. However, it might be partially developed at a trifling expense by excavating a dome-like recess above, and bareing the upper side of the fossil for three or four feet without disturbing any other part, so as to leave it *in situ*, after protecting it from atmospheric influence by a coat of varnish or other suitable material, it would then be an interesting lesson for the geological student. In excavating a deep open drain near to this sandstone cliff, and parallel to it, a great number of dark coloured concretions were thrown out, many of which contained a group of fossil shells as a core or centre. These were not of the type of any we have found in the Mountain Limestone, and probably some are of new species. Proceeding northwards we cross a beautiful plot of level land containing 128 acres, part of the new home farm of Holker Hall, a splendid example of what can be effected by skillful management, nearly the whole plot having been reclaimed from the sea. Its history is as follows: when the Ulverston and Lancaster Railway was constructed, ten or twelve years ago, it cut off by an embankment, between two rocky points on the shore of Morecambe Bay the tract of land referred to, then called Godderside Marsh, consisting of 48 acres of a rather poor salt marsh and 80 acres of sea sand, similar to the whole of the Ulverston Sands. This was the state in which G. Drewry, Esq., the principal steward of His Grace the Duke of Devonshire, took it in hand, but now it is the most beautiful plot of ground we have ever seen. He commenced by making roads across it where necessary, divided it into fields bounded by straight lines and right angles, and although fifteen or twenty acres each, there is none in which the breadth of the two ends differs to the extent of three inches. He then established a system of drainage which we will not attempt to describe, but these and all his subsequent works and experiments afford a valuable lesson for the practical farmer in every part of the kingdom.

Resuming our excursion northwards five or six hundred yards on the shore of the bay, we come to Capeshead cavern, long known as a fox hole, but in excavating the rock for railway purposes the workmen broke into a cavern upwards of seventy feet in length, and in some parts from fifteen to eighteen feet in height, varying in breadth from nine to twelve feet. The floor consisted of a hard indurated clayey material. His Grace the Duke of Devonshire, the owner, had the cavern partially but not entirely cleared out, when bones of the smaller carnivora with those of the hare, rabbit, &c., were found imbedded, but not in great quantity. We found firmly cemented in a mass of stalagmite the skull of a species of badger, with a remarkable crest-like ridge rising from the upper part of the cranium and terminating with a rather thin edge. When compared with the skull of the dog, the fox, and even with that of the common badger, although they all have a rudimentory ridge on the upper part of the skull, it was very different. The extent to which we can penetrate at present is only seventy feet, yet the cavern does not absolutely end there, but at this point a low opening by which a dog can make further progress into the hill of limestone rock may extend into another chamber probably larger and more interesting than that we have now under notice. Continuing in the same direction on the sea-shore, about three hundred yards, to Parkhead, is a good locality for Organic Remains, where we have obtained several splendid specimens of *Mitchliena grandis*, two or three of which are the best we have seen. There is an abundance of fossils, particularly the genus Cornulite, but many of them are embedded in the rock. For a considerable distance between Low and High Parkhead, there crops out amongst the shingle of the sea-shore a hard grey shale, from which storms and high tides detach many specimens of *Mitchliena grandis*, some entire and perfect. This is a rather scarce Mountain Limestone fossil, and was thought to be almost peculiar to Arnside. From Parkhead the fringe of limestone extends northwards by Low and High Frith, and Hazelhurst Point to the north end of Roudsea Wood, where it forms a junction with the great Clay-slate deposit of Furness, and throughout the last mentioned distance Organic Remains are not abundant in any part of its range. At this point the Mountain Limestone occupies only a small space on the surface, but theoretically it may be considered to stretch under Wreaks Moss eastwards to the Clay-slate ridge of Bigland scar, which courses southwards beyond the gardens of the Duke of Devonshire, at Holker. A small patch of limestone crops out near Stribers by the road side, apparently limiting its eastern range, indicating a fault parallel with the public road from Stribers to Holker, and it is evident the junction between the Clay-slate of Bigland scar and the long fringe of Mountain Limestone (detailed above) occurs under the moss allotments of Wreaks and Cartmel.

We have noticed the various geological phenomena of interest on the coast line of the parish of Cartmel. The interior of the parish will be given in exceedingly minute detail by the gentlemen of the geological survey of England, W. Talbot Aveline, Esq., F.G.S., and A. C. Grant Cameron, Esq., who, with competent knowledge of the science, and great industry and perseverance, have now almost completed the field work of the geological survey of Cartmel.

A geological ramble of some interest may be taken in an eastern direction commencing again from the Grange Hotel. After passing the Railway Station our road follows the line of a high cliff of limestone thirty or forty feet in height, cut through by the Ulverston and Lancaster Railway. The rock abounds with Mountain Limestone fossils, *Turbinola fungites* being especially abundant, but neither these nor any other species of Organic Remains are of much consequence, their sculpturing being very obscure, and most of them firmly embedded in the rock. Half a mile from Grange we cross the railway to Holme Island, one of the seats of Alexander Brogden, Esq., M.P., who, with more than the usual amount of liberality, admits the public two days in the week to visit and wander through all parts of this beautiful little island, which is truly an "Emerald set in the ring of the sea." He has recently, at a great expense, connected the island with the new road to Grange and the main land by a roadway over a solid embankment, so that it may be visited at any state of the tide. The whole of the island is composed of Mountain Limestone, large blocks of which are strewed on the east side, causing a tremendous uproar when ruffled by storms and high tides. Our traverse northwards to Castle Head is a pleasant walk of half a mile, and no geological student should neglect to visit this place. Castle Head has a remarkable sudden rise from the level plain, and consists of an immense mass of Silurian rock covered with wood, garden, and a mansion of considerable extent, eighty or ninety years ago the home of J. Wilkinson, Esq., the great Welsh iron master, but now the property of E. Mucklow, Esq. Many articles of antiquarian interest have been discovered at Castle Head, but as they are foreign to the object of this work, need not be noticed here. This strange and sudden uprising of rock offers a valuable lesson to the student. The south point of the rocky hill slopes quickly down to the level of the plain below and sinks into the ground, when it meets the general level, but the Silurian portion of the rock is seen no more on the surface in a southern direction until we find it repeated in North Wales. Eight or nine feet in height up the cliff, there occurs a deposit of Mountain Limestone which overlays the Clay-slate conformably, and in actual contact without either shale or conglomerate intercalated between them, and the junction or splice of the two formations is so perfect that a straight line may be drawn between them down to the earth, indeed it is the most perfect example of a junction between two different formations we can find in the whole district. This is the northernmost appearance of the Mountain Limestone in Cartmel, and it continues uninteruptedly to the extreme southern point of the parish, to that singular bluff of limestone rock—Humphrey Head Point. The foot of the remarkable hill of Rock at Castle Head, is washed by the river Winster, which for several miles has ever been the boundary between the counties of Westmorland and Lancaster, but at this point the river is now diverted so as to make its entrance into the sea upwards of a mile further to the east, at Meathop Point.

The great rock-cutting made in the construction of the Ulverston and Lancaster Railway, at Meathop Fell End, exposes a section of rock of some interest, dipping southeast at a very low angle, in some places nearly horizontally. There is a good quarry for strong plain work, such as engine beds, foundations, &c., but the stone is not suitable for sculpture, as it breaks with a splintery fracture similar to that of Old Hills Quarry, near Marton. Organic Remains are rather scarce, but in the southernmost field, almost immediately above the quarry, we may obtain a profusion of two or three different species of coral in nearly all of which the matrix or natural rock is weathered and gone, leaving the fossil entirely developed. *Calamapora megastoma* is especially abundant, many being strewed on the ground, some built into the fence walls, while others are turned up with the plough, most of them light and honeycombed, something like a bundle of rushes tied together, and when washed are beautiful cabinet specimens. We may now go forward to Arnside—the typical locality for *Michliena grandis*. From the shingle bed on the west shore of Arnside, most of the beautiful specimens of this fossil has been obtained. Numerous other species of Organic Remains may be obtained here, indeed we consider it one of the best localities in the whole district for obtaining Organic Remains in quantity.

In closing this scant notice of the geology of Cartmel, it is not necessary to attempt a formal description of the lovely village of Grange, as that is more the province of the poet and the landscape gardener than the geologist,— the airy and figurative for the former, the material and physical for the latter,— but its rare beauties must not be entirely left unnoticed. From its delightful situation and other natural advantages, Grange must, under all circumstances, be a place of fashionable resort, for nothing can prevent it, and when its charms are more extensively known and appreciated it is not improbable that a considerable portion of the eastern slope of Hampsfell will eventually become dotted with villas, and the permanent home of a portion of the "upper ten thousand." The most remarkable geological feature in the immediate vicinity of Grange is the immense natural terraces of limestone rock, rising perpendicularly to a considerable height above the surface, all containing Organic Remains characteristic of the Mountain Limestone. We have not noticed any rare species or new forms of life upon these interesting terraces, and that is so much the better, and we should deem it a sort of geological madness to destroy, injure by blasting, or in any other manner attempt to take away any part of the noble escarpments of rock which give an especial character, and a beauty entirely their own, to the landscape.

Grange is by no means a good geological field of study, the whole immediate neighbourhood consisting of one formation only, therefore offering no variety or change in the strata, yet some interesting Organic Remains may be obtained at the small quarries to the west of the village, and a walk over the south end of Hampsfell to the quarries near Allithwaite will reward us somewhat better. We have obtained there a few corals with their original colours unfaded, which when polished are beautiful specimens for the cabinet. In returning from this short ramble we must not neglect to visit the Hospice on the top of Hampsfell, and enjoy for a short time the splendid view from thence. We have no pretension to description of natural scenery or we would make an attempt from the summit of Hampsfell, and we certainly should not leave out Holme Island, with its sweet little Grecian Temple peeping out amongst the trees.

GEOLOGICAL RAMBLES IN THE LAKE DISTRICT.

Some time ago a communication was read before the Geological Society of London, by Professor Harkness, "On the Skiddaw Slate Series of the Lake District," in which he described it as being in some localities eminently fossiliferous and containing Organic Remains more or less through the whole of its range wherever he had made a minute survey. This announcement created a great interest in the geological world, the Skiddaw Slate having hitherto been considered almost devoid of such remains, with the exception of two or three found by Professor

Sedgwick, and as this formation is overlaid by the unfossiliferous Green Slate and Porphyry in the natural sequence of stratification, and this again by the Coniston Limestone hitherto considered the lowest fossiliferous stratification in the north of England, makes Professor Harkness's memoir especially interesting and valuable, and it will have the effect of drawing geologists from all parts of the country to the different localities described in the Professor's memoir as having yielded Organic Remains to his close and untiring scrutiny. Indeed, it has already had that effect upon ourselves and a geological friend, for we visited several interesting localities, and, although the weather was unfavourable for a ramble among the mountains, we could not resist the temptation to examine the matter for ourselves. It is not our intention to give many particulars of this journey, as they would be in no way interesting to the general public, but a slight sketch may be of service to geologists. We had arranged previously that our journey should be on foot—the only way in which a geological ramble can be effected—for our impression is that no man can know anything about a district unless he walks through it.

We left Ulverston on the 3rd of September, 1863, the rain falling heavily at the time, and we had no inducement to linger on the road, there being nothing of geological interest after leaving Hoad, until we passed Belle Isle and The Ferry, on Lake Windermere, the whole of the rock formation to this point being either Lower Ludlow Rock, or Upper and Lower Ireleth Slate, these stratifications blending imperceptibly into each other. The next stratification in the series of deposit, the Coniston Grit, which theoretically might be looked for immediately on passing Belle Isle, and about a mile beyond the island, nearly opposite the "Crier of Claife," we pass over that beautiful rock the Coniston Flag. This formation is slightly fossiliferous, and continues until we have passed Low Wood about a quarter of a mile, where it forms a junction with the Coniston Limestone. The Coniston Limestone is a long narrow band of dark blue rock, which shows at Beck Farm, in Millom, and thence takes a north-east direction for upwards of fifty miles. It enters Windermere at Poolwyke and issues about a quarter of a mile from Low Wood Inn, from thence it shows again at Troutbeck, Kentmere, and Longsleddale. If the water of the lake at this point were not more than ten or twelve feet deep, we should see this dark blue band of rock like a ribbon at the bottom, not more than forty or fifty yards wide. It is highly fossiliferous, and contains many rare and beautiful forms of ancient life, several of new species unknown to science. As before stated, this has hitherto been considered the lowest fossiliferous rock in the north of England. After crossing the Coniston Limestone, we enter on the Green Slate and Porphyry, which continues northwards through Ambleside, Grasmere, Wythburn, and a great part of the Vale of St. John, where it forms a junction with the Skiddaw Slate, the lowest stratified rock, resting immediately upon the granite. All the stratifications enumerated above, run parallel with the Coniston Limestone, and are deeper in the series, for although as we proceed northwards we are rising higher above the sea level, and still getting deeper and deeper in the stratification of the earth, until we come to the granite of the Skiddaw Forest, where there occurs an anticlinal axis, which somewhat reverses the dip, and soon brings to the surface the Mountain Limestone of Hesket and the Permian Sandstones of the Vale of Eden.

After leaving the head of Windermere, we proceeded through Ambleside without stopping, the rain falling very heavily indeed; but on we trudged, splash, splash, towards Grasmere. As might be expected, we were not very particular now about taking the cleanest part of the road, as we could not be much worse. After passing Rydal Mount, the seat of our lamented poet, we came upon a set of targets for rifle practice, in a beautiful glade almost surrounded by woods. It is very gratifying to find that even in this quiet mountain district our national defences are not neglected. We have always held that not only volunteers, but every man should learn the use of arms, and we do not acknowledge any one to be a perfect man unless he has some knowledge of the use of the rifle. By the time we reached Grasmere the rain had much abated, and the water ceased to run from our clothes, so that if it would remain fair they would begin to dry. At Grasmere we left the main road to visit the churchyard, in a quiet corner of which sleeps our Wordsworth; and close beside him poor Hartley Coleridge. After leaving Grasmere we soon began the long ascent to the summit of drainage at Raise Gap. Raise Gap, or Dunmail Raise, as it is sometimes called, is said to be the place upon which, in the year 945, the Saxon monarch, Edmund, defeated Dunmail, the last king of Cumberland. A large cairn on the left of the road is pointed out as the last resting place of the unfortunate Cumbrian. Sir Walter Scott alludes to the British bards tuning their lyres

"To Arthur's and Pendragon's praise,
And him who sleeps on Dunmail Raise."

Near to the Gap a small stream rises from the mountain on the east side of the road. This stream, which here runs southward, is the head water of the river Rothay, and flows through Grasmere and Rydal lakes, forming a junction with the Brathay five or six hundred yards before it enters Windermere Lake, and thence by the river Leven to the sea at Morecambe Bay. Immediately after passing the Gap there is another small stream, the fountain head of the river Greta, flowing northwards by Wythburn, through Thirlmere, and the Vale of St. John, entering

Derwent lake at Keswick; thence by the river Derwent through Bassenthwaite lake by Cockermouth, and entering the sea at Workington. After passing the Gap into Cumberland, we began the long descent into Wythburn—

> "Till on our course obliquely shone
> The narrow valley of St. John,
> Down sloping to the western sky,
> Where lingering sunbeams love to lie.
> Paled in by many a lofty hill,
> The narrow dale lay smooth and still,
> And down its verdant bosom led
> A winding brooklet found its bed."

The mountain range on both sides of the valley became more precipitous as we proceeded — Eagle Crag, Raven Crag, the Castle Rocks of St. John, &c. — and from their jagged and serrated sides it was easy to see that we were still on the Green Slate and Porphyry, and even after darkness set in we knew we had not yet reached the Skiddaw Slate, the object of our journey.

We arrived at Keswick two hours after dark, and remained there all night, the next morning proceeded to Barff, four-and-a-half miles from Keswick, on the road to Cockermouth. The hill of Barff rises abruptly from the road side; this portion of the mountain for several hundred feet in height is covered with an immense bed of shingle, the slope of which is as regular as the sides of any railway cutting, the gradient about one-and-a-half to one; it is very loose and slippery, difficult to ascend, but you can slide down without much effort, if you can keep upright. This shingle bed is Professor Harkness's favourite locality for Organic Remains.

We spent two days at Barff, and hunted it industriously the whole time, but we were rather disappointed, for it was two or three hours before we found a single fossil; at length our companion, Mr. J. P. Morris, gave the well known geological whoop, when we hastened to him, and it proved to be a rather good specimen, containing two or three entire figures of *Phillograpsus* (Hall), which we divided, one taking the fossil, the other the matrix in which it was embedded. After this, at long intervals, we succeeded in finding a few more of the same species, and at length our friend gave the whoop again, exhibiting what proved to be a very good specimen of *didymograpsus V. fractus*, N.S. We also found two ar three imperfect specimens of the *D. caducens*.

During the two or three days we were at Barff, the weather was very unfavourable, showers coming on without any warning. When a heavy rainfall began we laid down on the shingle, in a lump, covering ourselves with the umbrella, so that nothing else could be seen, and remained in this state — something like a snail in his shell — until the rain was over, when we arose and resumed work until the next shower made it necessary for us to creep into our shell again.

The next morning we proceeded to Braithwaite, and tried Braithwaite Brow, (another of Professor Harkness's localities) but without success. We then took to the mountains between the vale of Keswick and Crummock Lake, and in the long range between Causey Pike and Grassmoor, we found another of the Professor's favourite places where we intended to spend the whole day. This was a very exposed situation. It began to rain soon after we arrived, and we had no friendly rock that could afford us much shelter. We soon found this was better for Organic Remains than any place we had visited, but it was rather unpleasant to work in the rain, which began to fall very heavily, and continued through the whole day. We attempted to shelter two or three times against a low rock, with the umbrella over our head, but when night came on we were more thoroughly wet than our friend, who never attempted to shelter during the whole day; however, we could not laugh at each other, for when we arrived at Keswick, after a walk of five or six miles, there was not much that was dry about either of us. The fossils we found were all of different species from those we obtained at Barff, (a list of the known specimens will be given hereafter,) but some of them, no doubt, are new, therefore not named, or described, particularly a beautiful small fossil bi-valve shell, found by our friend, which rather resembles *lingula Davisii*, although much smaller. This, we believe, is the first, and at present the only true fossil ever discovered in the Skiddaw Slate.

The new crustacean, figured in the "Quarterly Geological Journal," for May, 1863, described and named by Mr. Salter, as *caryocaris Wrightii*, should have been called *caryocaris Grahamii*. Of this crustacean we found several specimens, also a few of *graptolites sagitarius*, *dipiograpsus pristis*, and one of that rare species *tetragrapsus*.

As our friend was obliged to be at home the next day, we felt we must accompany him, but made up our mind to come again alone when the weather was more settled, as we considered this merely a preliminary ramble. We put up at the "Lake Hotel" (Mr. Atkinson's), and they were very kind in assisting us to dry our clothes. The next morning we visited poor old Joseph Graham, a mineral dealer and shoemaker, in Keswick, who is not only the real discoverer of the new crustacean *caryocaris*, but of nearly all the different species of Organic Remains found in the Skiddaw Slate. We afterwards visited Mr. Charles Wright, a very intelligent geologist, who was exceedingly kind although we were entire strangers to him. We then commenced our journey home, where we arrived at ten o'clock the same night, without any adventure worth recording, after an absence of five days of unsettled weather.

GEOLOGISING UNDER DIFFICULTIES.

AFTER our return from the ramble already described, the weather improved a little and we began to prepare for another excursion to the same district alone. Our design was to make this one more for mining purposes than to seek for Organic Remains, and particularly to "prospect" for veins of lead, cobalt, and iron ore. We arranged to go by rail to Whitehaven, thence to strike inland by Lamplugh Cross, to Lowes Water, and the central mountains of the Lake District, to penetrate and examine the most solitary and unfrequented places. As we have had some little experience in sleeping on the mountains, viz., Dow Crags, Walney Scar, and on the top of Scawfell, we contemplated doing the same in this instance, as it is very unpleasant to travel to an inn at nightfall, and return to the same place in the morning. On all our Green Slate mountains there are many sheltered nooks among the rocks, so that any man in health may, during seven months in the year, take a few nights' lodgings in some of them, with comparative comfort, especially as he may have his choice of bedchambers, with sometimes a nice soft bunch of heather for a pillow. The Skiddaw Slate mountains do not afford the same shelter as the Green Slate and Porphyry, the principal part being soft, and wasting rapidly by atmospheric influence, so as to break down all precipitous rocks and leave the surface comparatively smooth and even. A good example of this is afforded by Skiddaw itself, which when viewed from Keswick, looks more like a lawn in front of a gentleman's seat than a mountain composed entirely of rock; yet places of shelter may be found even on mountains of the Skiddaw Slate formation.

We started from Lindal by the early train, without breakfast, on the 25th of September, 1863, a fine warm morning for the season, arriving at Whitehaven about half-past nine, and, after a ramble through the town, began our journey on foot to Lamplugh Cross. After a few miles' travel our old wallet began to feel heavy, and as we saw a cottage by the roadside at a little distance in front, we thought we would ask the inmates to boil the kettle for us. This was our first essay in that line, for we were now "doing the economical," and we did not care if they refused the favour, for we could light a fire by the roadside and make our own breakfast, as we had every necessary for that purpose in our wallet, and felt very independent indeed. However, before we reached the house, we overtook two farm servants, with horses and carts, returning from Whitehaven, with whom we kept company to the cottage, opposite to which was an old gate across the road, with a chain and padlock. Being a private road across Frizington Moor, this was a penny toll bar, kept by an old woman, who came stumping out with a pair of great clogs, a strange looking cap, and a bedgown which reached about six inches below the waist. When the carts had passed through she demanded her penny, and they said "Nay, Mary; we paid in the morning, and we will not pay again," But Mary insisted that they were liable, as they had two boxes in the carts. But they replied "No, Mary; thar's nought in t' boxes but corn for t'horses, and oor maister says he will nivver gie ye another penny, sea noo, Mary, tak ye that." The old woman dismissed them with a not very complimentary speech. During this little harmless quarrel about a penny, we stood in the road, waiting to speak to the old woman when she was at liberty.

When the men began to move on — the gate being still open — Mary turned round to us and said "Noo, me man, are ye gaun through?"

We replied, boldly, "No, Mary; I am going into your cottage, and you must boil the kettle for me for I have come sixty miles this morning, and have had no breakfast. I have some provisions in my wallet, and I will give you something for your trouble."

She replied, "Why, it's time ye had summat to eat, whativver ye be. Come in me man."

We saw that Mary was one of the real old Cumberland women, who are all dead, except herself and another or two, who may perhaps be found among the mountains where no railway whistle will ever reach them,

Mary's cottage was somewhat original; the floor was paved with large granite boulders, some of them three or four hundred pounds in weight, laid with the flattest side upwards, but there were large open spaces between them, filled with anything that came to hand. It could not be expected that the floor would be very level, and Mary had to turn her three-legged table round and round several times before she could find a place where it would stand, without upsetting the old broken jug in which my coffee was made; the cover was a broken cup which did very well. The old woman then put on the kettle, and stumped about in her clogs with right good will, and as soon as the coffee was ready she sat herself down in an old, carved, black, oak chair, by the fire side, and the following conversation passed between us. We will endeavour to give it word for word, as near as we can, and although it may not be very interesting, it has some bearing on geology. It also gives some shrewd, although rather uncourtly, remarks respecting the iron ore mines and miners of Furness. It will also serve to illustrate and explain other matters which will appear hereafter.

After resting a little while, the old woman looked at us and said, "Ye say ye hae come sixty miles this morning, whar may that be frae. Is it frae Forness?"

"Yes, Mary. I came by railway from Furness to Whitehaven, and then walked here."

The old woman shuffled about in her chair, and said, "Oh, dear, them nasty railways; they will nivver let us alain wi them? they're making ane frae Penrith to Cockermouth, and another frae Frizington to join it, and that will come net far frae here."

"Well, Mary, it may be better for you and better for the mines, for they can either ship the ore at Whitehaven, or send it by railway to the east country, where some of it is wanted."

"I doot that, me man; but surely this iron ore has become a terble bizness noo, and when I was a bit lassie, thare was nea ore, but a lile bit they carried i' sacks on their jackasses. But there was ane Mister Tulk, frae Lunnon, 'at hes set a' t' country crazy for ore," Then looking at us she said, "I suppose ye hae a lile bit of ore i' Forness?"

We were rather piqued at this, knowing that we had a deposit of ore in Furness that could not be equalled in Britain, and we replied "A lile bit did you say? I believe we have more iron ore in Furness than you have at Cleator, Frizington Lamplugh, and Knockmurton, all put together."

"That's a lee, hooivver," replied Mary, "for there's some o' yer Forness folk grubbing for ore baith at Lamplugh and Knockmurton just noo; and if ye hed sae muckle ore at hame, ye wad na come here for it."

This was quite a knock down blow, and we could say no more on the subject.

Before breakfast was ended the old woman began to be quite chatty, and she kept looking from us to the wallet, and from the wallet to us again; at last her womanly curiosity got the better of every other feeling, and she said, "I don't quite understand what ye are. Ye don't leak like a 'laker,' an' ye are nouther a hawker nor a beggar, what are ye? Whar's ye gaun?

We were much amused with the old woman's speech, and being willing to oblige her, said, "Well, I know you will laugh at me, and perhaps you will not believe me, nevertheless I will tell you the truth; and, as I said before, I come from Furness, my name is John Bolton, I am a

GEOLOGISING UNDER DIFFICULTIES.

poor man, but as you say, not a beggar. I am going to have a few days' ramble among your mountains; to go here and there, wherever my fancy may lead me; to eat when hungry, drink when dry, and when night comes on, lie down in some comfortable nook amongst the rocks and take my rest."

The old woman looked at us very earnestly and said, "D'ye say ye're gaun to sleep on t' mountin ta neet?"

"No, Mary, not to-night, but very likely I shall to-morrow night."

She replied rather sharply, "Why that is a bigger lee 'an that other ye tell'd me, when ye said ye had a deeal o' iron ore i' Forness, an' at t' saem time, Forness folks come seeking for it here."

"Now, Mary, that is twice you have told me I was lying; now, if you had been a great gentleman, instead of a kind-hearted old woman, and I had been a gentleman also, we should have had to fight a duel, and perhaps one of us would have been killed; but we can settle it better than that, for I know that you did not intend to insult me, therefore I excuse you, as I know that you Cumberland folks have a way of joking and bantering each other, different to anything we have with us, for in Furness, whoever has the least pretension to decency, is supposed to speak the truth, and I speak the truth now, to you."

We thought we had made a very grand speech, and quite finished the old woman, but it was a mistake, for she replied, "Why, then, if ye are speaking t' truth, as sure as my name is Mary Hasting, ye're crack'd,—that's a'; for ta think of an auld man like ye, talking aboot sleeping under a rock, on t' mountins, at t' latter end o' Septemmer, and i' sic weather as this. Why, ye had better drown yoursel i' Crummock Watter before ye come there." After waiting a moment, she said, "Why, ye don't leak like ane that is wrang in his heed, an' ye talk farrantly enough; hae ye ony family me man, an' dea they knaw what ye are gaun ta dea?"

"Yes, Mary, I have a family, and my wife put up all the provisions in my wallet, except two loaves I bought at Whitehaven."

"Why, then, yer wife is crack'd teah, and somebody should be appointed ta tak care o' ye beath; but tell ma what ye want on oor mountins, an' what d'ye expect ta finnd?"

"Well, you see Mary, there was a great man came from Ireland, they call Professor Harkness, and he found some stones with curious markings on them, and he has written a book about them, and I thought I would try to find some of the same, and, perhaps, some such as the Professor had never seen; besides, I might happen to find a little vein of lead, or iron ore."

To this, old Mary immediately replied, "Why, noo, me man, ye hae let t' cat oot o' t' bag, ye dinna care muckle aboot them stanes, it's all plain ta me noo; why, ye're just sent by them Forness folks to spy about, an' try to finnd some mair iron ore for them, for I hear they are no doing sa weel, nouther at Lamplugh nor Knockmurton. Why, why, it's reet enough on yer part, if ye're weel paid for it; but it's a queer thing that them Forness folks can't be content wi their awn, if they hae so muckle o' it."

So the old woman settled our business at once, and with a pertinacity characteristic of her age and sex, she maintained without any evidence whatever, that such was the object of our ramble to the mountains; and so the matter ended.

It was evident this ancient dame could not understand that there may be inducements to follow some scientific pursuit quite independent either of profit or fame, and to contend with difficulties and dangers entirely for the love of science itself.

After a while, our entertainer resumed, "I would like

ta knaw hoo ye are provided wi meat; hoo ye will cook it; an' hoo ye will contrive ta sleep?" Being willing to satisfy her in this also, we took out our wallet, and emptied it on a long black oak table, that stood before the window, and said, "you see I have plenty of bread, and I have two or three pounds of boiled ham in this newspaper, some ground coffee in this small canister, and some butter in this canister, and plenty of sugar in this paper, and here is a tin can that will hold half a pint; this is a very important utensil, for it is my kettle, my coffee-pot, and also my coffee-cup. As for sleeping, you see I have a top-coat on my back, and I have an old wallet inside of this new one; this will be my bed, besides I have a box of matches in my waistcoat pocket, to keep them dry, and I think I am not badly provided."

The old body laughed outright, and said, "Why, ye Forness folk have na a' t' wit in the world; now this wallet is only aboot half as lang as yersel, what will ye dea wi' yer great lang legs? — ye hae forgitten them, I'm thinking."

Rather annoyed at the old woman's mirth, we replied, mildly, "No, Mary, I have not forgotten my 'lang legs,' and I think I can take care of them also. Perhaps you have seen a little animal called an urchin, or hedgehog, which rolls itself up in a lump like a football; I must endeavour to roll myself in the same manner, and then the wallet will be long enough."

"Why, thar is another thing waar than that. Thar is nea a yerd o' dry ground on a' oor mountins. Will ya lay doon on the damp ground, or, maybe, in the wet an' mire?"

"I will tell you how I have done in a case like that, and it will be a lesson for you when you sleep on the mountains."

"Nay, me man, but ye're joking, noo; but I hope I'll

nivver lay on t' ground until they lay me in Lamplugh Churchyard, whar o' me kith an' kin are laid. Hooivver, ye ma let me knaw."

"Well, you see, Mary, five of us agreed to sleep on the top of Scawfell, and there were the ruins of a place made by the Ordnance surveyors. It was about as high as your table, but it was very damp, so we laid flat and dry stones side by side, where we intended to sleep, and made it something like your rough floor. Whenever I attempt to sleep on the mountains I will do the same; but, indeed, Mary, there is nothing so very terrible in it after all; and, besides when we patronise the 'Rock Hotel,' it is some comfort to think that there is no landlady to come to you in the morning, with a little bit of paper in her hand, with so much for supper, so much for breakfast, so much for bed, and so much for attendance; you see nothing of this. All you have to do is to return thanks to the Almighty for watching over and preserving you through the night, and go your way with thankfulness and a cheerful heart."

"Ye draa a fine pictur o' yer 'Rock Hotel,' as ye ca it, but ye'll finnd it oot yet. Noo, dea ye tak onything to keep tha cauld oot, for I see nea flask as they ca it?"

"No, Mary, I never drink anything strong?"

"Oh! I see; ye're a teetotaller."

"No, I am not a teetotaller. I will never sign their pledge, and bring my mind into bondage about it, and yet I have not tasted ale, porter, wine, or spirits, for a lifetime almost, and I do not feel the want of it; indeed, I have almost forgotten the taste of all of them. And now, Mary, I have told you more about myself than I intended, you must tell me something of your life."

"Why, my tale is seean telt; thar has na been mony changes in it, for I hae leeved here for mair than sixty years; for, ye see, my folks wor farmers at Lamplugh, an'

I went to sarvice a Frizington Ha', when I was a bit lassie, an' I was married fra thaar when I was a varra young woman, an' me husband brought ma here to this hoose. Why, indeed, it wasna a hoose, I may say, for me husband an' me ameast made it oorsels, an' mony a weary day we hed, an' I goul'd, an' I goul'd like a bairn, for it was a unco change for me ; but we hae browt up a family here, an' me husband was varra kind ta me. He has left ma twa year sen, an' I mun sune follow him ; but I mun wait patiently the Lord's time. Hooivver, I will nivver leave this hoose till they carry me oot to Lamplugh Church, for there isna a stane in o' t' flure but brings back some auld thing to me mind." And the poor old creature was affected to tears.

"Well, Mary, you have everything very clean. Have you all the work to do yourself?"

"Nay, I have a daughter at draas ta ma, an' she cleans and tidys up for ma, an' is varra kind to her auld mither, or I couldn't dea at a'."

We could not prevail on our hostess to take anything for the trouble we had given her, nor even to have a cup of coffee with us. So we parted at the cottage door, with kindly greetings, and throwing our wallet over our shoulder, we went on our way, not without a feeling of sorrow for this lonely old woman. We had now travelled two or three miles on the Mountain Limestone, and seen a great many large granite boulders, similar to those with which Mary Hasting's cottage was flagged or paved (for it was a mixture of both). Some more than half a ton each, were lying by the road side at several places between Hensingham and Lamplugh, they were covered with a white crust, due to decomposition, and looked like chalk. It is rather singular that we did not observe any other kind of boulder than this, especially as there is no granite rock *in situ* nearer than Skiddaw Forest, or Eskdale. The Mountain Limestone continues through Lamplugh, by Lamplugh Hall, and forms a junction with the Skiddaw Slate (on the moor, about a mile beyond Lamplugh Hall), without any intervening formation, therefore, the Mountain Limestone, the highest stratification, except one, in the Lake district, comes down on the Skiddaw Slate, the lowest stratified rock in the north of England; and all the different stratifications, enumerated as occurring between Ulverston and Keswick, are wanting in this part of Cumberland — viz., Lower Ludlow Rock, Upper and Lower Ireleth Slate, Coniston Grit, Coniston Flag, Coniston Limestone, and the Green Slate and Porphyry. At Lamplugh Cross, we left the road a little way, to see the iron ore mines of Schneider, Hannay, and Co. There are two shafts sunk, but very little doing. We only saw one miner, and his account of the works was not very flattering.

About two miles beyond Lamplugh (on the moor already alluded to) there is a small quarry by the road side, which has been excavated for road material. This is the first place where the Skiddaw Slate is seen *in situ*; there was a small quantity of *debris* about it, in which we found *Graptolites sagittarius*, and a portion of another fossil, which we could not make out. This would be a favourable place for Organic Remains, if well opened out by excavation. This high moor may be considered an extension northwards of Blake Fell, where Professor Sedgwick (that first of geologists) many years since obtained *palæochorda minor*, *chondritus acutangulus*, &c., therefore, we claim for Professor Sedgwick the credit of being the first, and, after him, poor old Joseph Graham, of Keswick, who have found Organic Remains in the Skiddaw Slate. Certainly, the recent extensive and most minute surveys of Professor Harkness, have added much to our knowledge of the Skiddaw Slates of the Lake district, and the thanks of all geologists are due to him for his important and valuable memoir on the subject. We have mentioned the above principally because the name of another individual has been brought prominently before the public, as having discovered several new species of *graptolites* in the Skiddaw Slate, whereas, we have it from more than one authority in Keswick (where this individual is well known), that he never found a fossil in the Skiddaw Slate in his life, but that he purchased the whole of those for which he claims that credit. It is with some reluctance we write this, and certainly we should not have recorded this unworthy act, had not the individual above alluded to courted publicity in the matter, and unblushingly taken credit to himself which he well knew was due to another.

About a mile from this small quarry, we leave the moor, the road turning abruptly nearly due south, by Water-end, thence on the east side of Lowes Water, the hills abutting the road, in some places having their slopes covered with shingle. We did not succeed in finding anything here; but as night was coming on, we did not examine them with sufficient care, but hurried on to Lowes Water Church, near to which there is a comfortable inn, (the "Hare and Hounds," Henry Pearson), where we put up for the night. After laying by our wallet, we repaired to the kitchen, where there were three sheep farmers, conversing on sheep and shepherding. This was just the company we wanted, as we thought to make some inquiries respecting the great and comparatively unknown block of mountains between Crummock Water and Keswick, which it was our intention to penetrate and explore in the morning; but we were surprised to hear that none of these farmers had ever been over them in their lives, although they had been on all the other mountains within many miles. Even the landlord, who was a great hunter, knew nothing about them. They all knew it was twelve miles from Lowes Water Church to Keswick, to go round by Buttermere and Newlands, and believed it would be considerably less if they could go direct up the rough mountain valley, between Grassmoor and Whiteside. Their knowledge of the locality did not extend beyond this, and we were strongly advised not to attempt to go that way, for we would find it rougher than we thought; but these shepherd farmers did not know that it was the roughness we wanted, and as for the direction we ought to travel to reach either Causey Pike or the village of Braithwaite, we were as well acquainted with it as any shepherd in Cumberland.

On the morning of September 26th, we commenced our journey, the mist hanging more than half way down Grassmoor and Whiteside, with showers now and again, and it continued so for four or five hours. After walking about a mile, we came to the foot of Grassmoor, the west side of which is very steep, and covered with screes, or shingle, almost from top to bottom. We did not attempt to examine this face of the mountain, but proceeded by a large beck, coming down between it and Whiteside, which was flooded, and although we could not cross it without wetting ourselves, and it was not quite so rough on the other side, yet we had to keep to the Grassmoor side altogether, which soon began to be very rough indeed. We had not only to tack, like a ship against the wind, but sometimes to go back a little way, and then try higher up the mountain side, and we found, after two hours of hard labour, we had not made half a mile forward, but we took it very patiently, and toiled on, sometimes over a bed of shingle, wet and slippery; sometimes on the surface of the rock, sloping quickly down to the beck; but still we were making some progress, and at length were able to proceed steadily, though at a slow pace. It did not surprise us now that the people at Lowes Water Inn had never been over these mountains, for it would be no easy task to any one. About one o'clock the weather cleared up, and the mist lifted, so that we could see the tops of the mountains, and the road on the hill side had so much improved that we could travel with safety and comfort, and we were cheered by finding a rather good specimen of the new crustacean *caryocaris Wrightii*. It was evident that the worst part of the journey was past, the Grassmoor side of the valley being free from shingle, and the hill side covered with soil a considerable way up, but it was crowned with a steep escarpment of rock, which we purposed to examine another day, if the weather would permit. After three or four miles, this valley ends suddenly, and is cut off by a steep and narrow ridge, which bridges it across the valley from mountain to mountain, like an immense railway embankment. This ridge separates the valley from that of Coldale, which may be considered a continuation of the former, and in the same direction; it is also the summit of drainage, the water falling both ways from this point. The large beck, already mentioned, which divides Grassmoor from Whiteside, flows westward, and falls into the river Cocker, near Scale Hill. On the east side of the ridge, a small runner (which soon becomes a large beck) falls from a higher part, flows to the east through Coldale and the village of Braithwaite, and falls into the river Derwent, in the Vale of Keswick. We have been more particular respecting these two valleys, because Professor Harkness, in speaking of Coldale Valley, says, "This valley seems to indicate the position of a well-marked anticlinal, extending in a W.S.W. direction through the Skiddaw Slates, which lie to the westward."

This is not only true, but it is also true in a greater degree, perhaps, than the Professor anticipated, for there is not only an anticlinal through the whole of the valley of Coldale, but it is also continued in the same direction between Grassmoor and Whiteside, and thus forms an immense crack entirely through this great block of mountains;

so that, if it were not for the ridge already mentioned, we would be able to see from about Lowes Water Church through the whole length of both valleys to Keswick. This ridge, although covered with soil and comparatively smooth and level, is rock under the surface, and forms an anticlinal axis at right angles to the one already described. The ridge is called Coldale Hawes, it is well seen from Coldale and the village of Braithwaite. From this point there is an extensive and beautiful view — Causey Pike in the distance, to visit which was one of the principal objects of this ramble. We now had a feeling of the most perfect independence, and it was almost a matter of indifference where we should sleep; but we had made up our mind not to leave the mountains that night, and unless the weather was very rough indeed, not for two or three nights more, as we had sufficient provisions for that time, and lodgings at free cost. But, as we said before, although some of the Skiddaw Slate mountains have many rough and serrated steeps, there are few overhanging and sheltered nooks to afford a snug retreat from the storm, like those in the Green Slate and Porphyry, and, as will hereafter appear, we did not make a good choice, for although we were sheltered from the wind, we were not from the rain. After leaving Coldale Hawes, which in itself is rather a geological puzzle, we proceeded in the best way we could in the direction of Causey Pike, where we had been two or three weeks before. But as the slope of this mountain range was here covered with very rough shingle, we had to descend almost to the bottom of the valley, and rise again when it became better travelling above. When we came within half a mile of the scene of our former labours, we passed the old cobalt mines, that had been given up many years before. We were not within a quarter of a mile of them, yet we could see the adit or level in the side of the mountain, and soon after this we arrived at the end of our day's journey.

GEOLOGISING UNDER DIFFICULTIES. 199

The first thing was to look out for a lodging, and we began to think we should not have passed the old mines, however, we would not go back again, but there being no friendly nooks amongst the rocks, we went forward to an old sheep-fold we had seen in a former ramble. This was a square inclosure, the walls being about five feet high, and the ground wet and miry all over. There was a sort of gateway on the east side, and an old gate or hack was lying on the ground, formed of four bars about a yard long, roughly nailed together. We carried this to the west or sheltered side of the inclosure, to be used as a bedstead, and not a bad one either, only it was about three feet too short; but we were thankful for it, as it would keep us out of the wet and mire which covered the whole place.

Perhaps it may afford some amusement to our friends, to know how we extemporised all that was necessary for our comfort during the night. We began by taking out of our wallet all that we had enumerated before; first the piece of ham in the newspaper, which we put in a large hole in the wall, our principal store room; we put the coffee canister in another hole, the butter in another, the bread and sugar in another; so we thought we had a respectable show, but there was no one but ourselves to see it. The two Whitehaven loaves were kept in the wallet to serve for a pillow when we went to bed. We then went in search of something to kindle a fire with, and brought some pieces of heather and a few green brackens. We now began to prepare supper, which was also our dinner, for we had eaten nothing all day; but as the sheep-fold was high up the hill side, there was nothing but bog-water within a considerable distance, so we filled the half-pint can with this, put the coffee and sugar into it cold, and set it down whilst we lighted the fire; but as the wind was rather strong, it made the smoke whiffle about so that we could not come near it, therefore we waited until it had burned out, when we put the can on the red ashes, and it was soon hot, though it did not boil. We then sat down on the little gate (in the same way that a tailor sits at his work) and ate our supper with a thankful heart. We put by the remainder of our provisions in the wall the same as before. It was now dark, and we were where we must remain the whole night; it was not a situation altogether new to us, but we had never before been entirely left alone, yet the lonesomeness of our situation was a luxury of no common kind, and well calculated to induce feelings of seriousness and devotion, and to lift the heart and mind from the things of this life to the Almighty Creator and Ruler of the Universe.

We were now, really and truly, on the mountains, far from the dwellings of man, and from the voice of every living thing; there was not even the bleating of a sheep to disturb our meditations, but there was the sighing of the wind amongst the heath and rocks above, and the rushing of the water below, which in Holy Writ has been called the Voice of God. It seemed that we had thus presumptuously placed ourselves almost in the presence of the Almighty, and we felt intensely our utter insignificance and worthlessness. Serious thoughts will always visit us on the mountains — we will not say intrude upon us, for serious thoughts should always be welcome — and we have always held that the mountains were made almost expressly to inspire and cherish devotional feelings, that they are churches made by God's own hand, and that no sermon can be so effective as one preached on a mountain. Our Saviour himself preached on a mountain, and prayed on a mountain; He was transfigured on a mountain, and many of the principal events of His life were intimately associated with the mountains. It has been a matter of surprise to us that none of our popular preachers (whose churches and chapels are not large enough to contain their congregations) never, even when an opportunity offers, should wish to preach on the mountains.

We had laid the little gate close to the wall, and as we had no pots to wash, we made our bed,—that is, stretching our two wallets upon it, we laid down, with our head on the two loaves for a pillow, and our back against the wall, commending ourselves to Him who watches over all. As our bedstead was too short, and we did not find it so easy a matter to roll ourselves up like a hedgehog, as we had said to the old woman, yet we could manage to shorten ourselves, and lie something like a dog or cat. This position at length became painful, and it was a relief to stretch ourselves out for a little while; but our legs, from the knees to the feet, hung over into the wet and the mire, therefore we got up and laid stones to raise them level with the bedstead, so that we could occasionally change our position. However, we thought old Mary was not very far wrong when she asked us "what we wad dea wi' our great lang legs." After lengthening our bedstead in the manner described above, we laid down again. The bars of the little gate felt rather sharp, and had it not been dark, we should have filled our old wallet with heather and brackens, and made it a most luxurious bed indeed. However, it was not bad after all, and we soon afterwards went to sleep. When we awoke, it was some little time before we knew our whereabouts; we could see there was no roof over head, and the full moon was high above us, now wading through dense clouds, and again shining with renewed splendour; the mountain tops were clearly defined, particularly Causey Pike, which was directly in front, and from the direction of the moon with respect to this mountain, we knew it was past midnight. We had now recovered full consciousness of our situation, and we remained awake a long time enjoying the scene. There was a solemnity in the very loneliness of the situation, which had a charm in itself, though of a melancholy chaaracter, and well calculated to bring the heart to commune within itself, and to review with seriousness the different scenes and events of a long life. At length we fell asleep again, and when we awoke it was daylight. We kept our bed about an hour longer, and arose very much refreshed.

It was now Sunday morning, September 27th. We intended this principally to be a day of rest, to begin in earnest on Monday morning. It was not necessary to waste much time in dressing; we had only to shake ourselves a little, and begin to prepare breakfast. The first thing was to go and fill our little can with water at the bog, put coffee and sugar in it, set it down, gather heather and brackens for a fire, put the can on the ashes as before, and as soon as it was hot, our breakfast was ready. We did not altogether like eating alone, and we thought if Professors Harkness, Sedgwick, Phillips, Ramsay, or King, should come that way, although we could not offer them a cup, we could give them a can of good bog-water coffee, with a small slice of ham, and we believe none of these distinguished geologists (who are all workers on the mountains) would have been above it, for men of science are no epicures, and most of them have already called upon us; Professor Sedgwick and Dr. Gough once condescended, with the greatest cheerfulness, to take a cup of coffee with us in our own cottage, before it was plastered — even before the floors were laid, and when grass was growing in the parlour. However, they did not make their appearance, so we took our breakfast alone, sitting on the little gate, with our back against the wall, and we made a very comfortable meal. The wind had been strong all the morning from the south-west, and wild clouds scudded rapidly between us and Causey Pike, far below the summit of that mountain, and before breakfast was finished it came on a smart shower, but was soon over. It was not long, however, before it thickened on all sides, and it began to rain in earnest, the wind increasing to a storm; but from the strength of the wind we did not think the rain would continue long, and we should have been thoroughly wet before we could have reached the old cobalt mine, otherwise we would have adopted a scheme (which it is not necessary to describe) to have enabled us to keep all our clothes dry except our great coat. Here we sat on the gate, crouched up, with the umbrella over our head, but it would not cover us entirely, so that there was a stream of water running down us either on one side or the other, and we found also that water was running down the inside of the wall and down our back. At length we became wet all over, and there was no sign of the rain ceasing, as hour after hour it poured down without any intermission. We now found we had made a great blunder in not taking up our lodgings in the old cobalt mine, but it was too late now, so we must profit by the lesson, and make a better choice the next time. As we were in the vicinity of the place we intended to work on the morrow, with no means of drying our clothes, it would be very unpleasant to work all the next day, and as we had originally intended, for two or three days more, we reluctantly concluded to go, when the storm ceased, to the village of Braithwaite, where we could dry our clothes, put up for the night, and come back again in the morning. We had hoped the rain would cease about twelve or one o'clock, but the storm increased, the rain becoming heavier, and about that time a stream of water burst through the north wall of the inclosure, (that being the higher side), and ran under the bed. The gate-bedstead, however, kept us out of it; though it did not matter, for we could not be much wetter, still we did not like to be intruded upon by this new river, so with our geological hammer we made a trench, and turned it another way. The work was rather a relief to us, as we were cramped with sitting in one position, and we now stretched ourselves up a little. Two o'clock came, still no sign of the storm ceasing; three, four, five,— it was now worse if possible, and as we had three or four miles to travel, we should have no time to spare to get off the mountains into the cultivated country before dark, so we took up one of our wallets, and left everything else, pushing the umbrella under the bedstead to keep it from being blown away. We then left the inclosure, and boldly stepped on the plain hill side. We now found the value of the wall as a shelter from the wind, for it was with the greatest difficulty we could stand, although our road was in the direction the wind was blowing, and we could not possibly have gone against it. We had proceeded about one hundred yards when we came to the remains of the road or track made by the miners when the works were in progress, but there was not much road now. The track was scratched out of the side of the hill for about two miles, and lay all the way close to the brink of the valley, or rather gorge, for it was too steep to be called a valley,— it was more like an immense railway cutting,—a large beck ran at the bottom of it, which was swelled by the rain to a roaring flood, white as snow.

We have been rather particular in this description, otherwise the predicament we were soon after placed in could not be understood by those who have not seen the place. Although it was difficult to stand near the inclosure, we found we had not got into the real current of the storm down this gorge, where it was confined by a mountain on each side, which intensified its effect to such a degree that we believe no man could stand before it. Soon after we came to the road we heard a sound behind us, not like that of the wind, but more like thunder at a great distance. We had not gone twenty yards before a terrific gust of wind swept us down instantly. It was at the turn of the road so that we fell against the hill side, lying flat on our face, we held on by the heather until it had gone by — had we been in a straight part of the road at the time, we should have fallen on the other side and come to an untimely end. We laid a few seconds until the gust had swept by, then rising carefully, and with difficulty, we attempted to return to the inclosure, but we found that to be impossible, therefore we must go forward at all hazards. We accordingly turned in that direction and had not advanced more than thirty or forty yards before we heard the hurricane coming behind us, so we instantly fell flat on our face, and held by anything we could get hold of until it had passed over. We had now gained some little experience, and as the sounds of these tremendous gusts gave notice of their approach, we always laid down flat on the ground, sometimes holding by the heather, sometimes by scratching with our hands into the shingle or any roughness on the path. We found we were making very little progress in our journey down the valley. Sometimes before one gust had gone by we heard another coming, therefore we laid still until that had passed also, when we got up and made the best of our way again. The rain during all this time was pouring down in torrents, but that gave us no concern. We had still a long distance to travel and we could not help regarding it as rather a perilous journey, but there was no help for it now, as we could not get out of the gorge, so we toiled on, and after several mishaps, which we need not mention, (as it is not our intention to make this a sensation article, although there was sufficient of true matter for the purpose,) at the end of three hours, we turned round a spur of the mountain, towards Braithwaite. We were soon in some degree sheltered by the mountain itself, and out of the line of the storm's greatest fury, notwithstanding this, the wind had not abated in the least, thus showing that local conditions have a most powerful effect either in increasing the power of storms or in lessening them.

We have now lying before us a letter from Mr. John Jackson, of Thrush Bank, Lowes Water, dated December 7th, 1863, which gives some account of a fatal accident, occasioned by a hurricane similar to the one we have described.*

After turning round this bluff we were able to cope with the wind, and proceeded northwards, but we were still high up the side of the mountain,—as it was the straightest road to Braithwaite. In this matter also we had made a bad choice, we would have saved ourselves some trouble if we had gone down into the public road, at the low end of the Newlands Valley, and from thence to Braithwaite. After travelling a considerable distance, we came to difficult places, which made it necessary to rise up over the top of the hill, and make for High Coldale, as we knew there was a road from Coldale to Braithwaite, but there was a valley between us and Coldale, with a roaring

* On Friday, the 15th of December, 1848, John Rigg and his two sons, all of them quarrymen employed at Honister Crag slate quarry, were overtaken by a violent storm of wind while in a defile of these mountains, and (like ourselves) at the approach of a gust of wind they threw themselves down flat on the ground, to prevent their being hurled down the ravine, but sad to relate, notwithstanding this precaution, and although the father lay in the middle, held down by his two sons, when the storm reached them the poor old man was torn from their grasp, hurled amongst the rocks, and almost dashed to pieces, his body being fearfully mutilated by the fury of that terrible blast. This frightful occurrence is well remembered by many people in Cumberland, particularly by those living in the neighbourhood of Lowes Water, and Buttermere. A full account of the extraordinary accident was given in all the local papers at the time of its occurrence. Mr. Jackson further states that "these furious gusts of whirlwind frequently take up great quantities of shiver and stones (as large as those commonly used to repair roads with), which fall like hail, to the annoyance and often destruction of sheep and cattle. Shepherds and workmen engaged in quarrying this far-famed slate, have frequent opportunities of witnessing these destructive blasts, and see them at a distance."

stream which we could not cross. This stream is a branch of Braithwaite Beck, so we followed it down towards the village, thinking to find a bridge, but there was neither a bridge nor a place where it could be crossed with safety; at length, when we came near the village, we were stopped entirely in that direction and had to go partly round the south side of the village, through some rough fields, and enter it by the public road at last. It had now been dark for upwards of two hours, and those trifling mishaps were rather annoying, but there was nothing dangerous after leaving the storm-valley, where our whole thoughts were concentrated on our own personal safety. During the latter part of this unpleasant walk, the old woman's words came into our mind: "Why, ye draa a varra fine pictur o' yer Rock Hotel, as ye ca it, but ye will finnd it oot yet." Certainly we had found something out—we had found it was the want of our Rock Hotel that had perilled our life, and made us now look something like a drowned man, and forced off the mountains. But this was all the result of an error in judgment, or rather a most unaccountable blunder in not making choice of the old cobalt mine for our home, where we should have been sheltered from the storm, instead of the sheepfold, where we were exposed to it all; besides this was an exceptional case, for in a conversation with the landlord of the inn where we stopped that night, he said there had been no storm like it for many years, for in about nine hours the flood had joined Derwent and Bassenthwaite Lakes into one, although when at their usual summer level they are about three miles from each other. After it became somewhat dark we began to flounder, stumble, and splash down the wet hill side which formed the south side of the valley of Coldale, and had now and again a little fall, yet this did not prevent us from attempting to amuse ourselves with the thought of meeting with one of our friends (which was not very likely) who had *not* been bitten with the "geological mad dog," and who also knew the object of our journey, and what we had passed through the last two days. We had made up a speech for him something like this. "Oh, John, you are an old fool! thus to subject yourself to storms of wind and rain; to exhaust your strength; to labour and toil, and even to peril your life. And what is it for? Why, for a lot of old stones with a few scratches on them, the best of which are not worth a farthing a bushel; and here you have been two days and one night on this block of mountains, and have just escaped from a situation of danger which it is fearful to contemplate, and your condition at this moment is not very enviable. But let this be a warning; you have already seen seventy-three summers, your thoughts should be on something else than rambling among these mountains. However, I hope this day's work will cure you, and you will go home in the morning and be content, for you have had enough of the storm for a lifetime." If the above speech had been real instead of imaginary, our reply would have been something like the following. "Well, what you say is very reasonable, but notwithstanding that, if we live, and be well, we shall come again to these mountains in the morning,—unless the storm should continue—for another day, for that was our errand here, and we believe nothing can cure us, for there is a charm and a fascination which draws us, and will continue to draw us to the mountains while we have strength for the work. With regard to those other thoughts which you allude to, as we have before said, the mountains are churches built by God's own hand, and more calculated to inspire those other thoughts than the most splendid cathedral ever built by man. Man can do many wonderful things, but God only can make a mountain."

When we reached the village of Braithwaite, people were anxiously watching to see if the flood would sweep away the town bridge; but we hastened on to the Royal Oak, a small inn kept by Isaac Fearn, a kind and intelligent old man, with several grown-up sons and daughters, who were all very kind. Our host ordered us to strip, and he brought us a suit of his clothes to put on, while they began to prepare to dry our own, by heaping logs of wood on the two kitchen fires. After we had "donned" the landlord's garments, we all sat round the fire, each having some article of clothing to dry. We began to be very comfortable, the old man relating some of his adventures on the mountains, and we told a little of ours in turn. When we expressed our determination to repair to the same part of the mountains in the morning, and intended to sleep on them again, he would not let us rest until we promised to come back to the Royal Oak at night. In the morning, as soon as breakfast was over, we set out for Causey Pike, the weather being rather wild and showery. We made direct for the sheepfold, as our two hammers and chisel were there. We found everything as we had left them the day before, the canister in the wall, being water-tight, was no worse; the ham only a little wet, and the newspaper was washed to pieces; the loaves had gone to paste again, but that was a trifle, as we had promised to sleep at the Royal Oak again; but in future we will make no promises of that sort, but leave ourselves at liberty to patronise the "Rock Hotel" if it suits us. We had also left a shirt and pair of drawers at the sheepfold, which we wrung as dry as we could, and spread on the rocks to dry, putting large stones on them to prevent the wind from blowing them away, so that it looked rather like washing day. We had plenty of drying ground and nothing to pay for it. Hitherto our time and labour had been wasted, but we now began to work in earnest, and found several good specimens of the new crustacean *caryocaris Wrightii*, and *diplograpsus pristis;* also *graptolites sagittarius*,

o

G. tenuis, and two or three other species which we have not examined minutely; also one specimen of *tetragrapsus, dichograpsus aranea,* (Salter). We also found a beautiful small fossil bi-valve shell, of the same species as the one we found in our first ramble, and very clearly developed; we also obtained the matrix, in which it was imbedded, in a very perfect state. But the most important fossil we have ever found in the Skiddaw Slate, was a new species of *Æglina*, noticed in the Geological Notes and Queries, in the following terms: — "A new species of *Æglina*, a remarkable genus of Lower Silurian Trilobites, rarely met with in this country, has reached me from Mr. Bolton, of Ulverston, who obtained it from the Skiddaw Slates. The beautiful crescent-shaped eyes of this trilobite are well shown in the specimen. This specimen was shown by me to J. W. Salter, who declared it to be the *finest specimen* he had ever seen."—G. S. Roberts, F.G.S., Hon. Sec. A.S.L.

This locality was the best we had visited. Organic Remains were not abundant, but after four or five hours of hard working we had made a considerable collection. During the afternoon there were two or three showers of rain, and as our shirt and drawers, spread on the rocks, were almost dry, and we did not want them wetting again, we had to run when a shower was coming on and fold them up until it was over, when we spread them out again as any other old washer-woman would have done. The day still continued wild; at length it darkened suddenly, and we had heard thunder at a great distance for some time, but now the storm was coming rapidly, and being above it we could see its breadth clearly defined, like a long black belt, not more than a mile in breadth. It came in the direction of north-west, and south-east, and passed about half a mile north of the place where we stood. We could see clearly that the sun was shining beyond it, and it would have been no difficult task for the raven or any of our mountain birds to fly from the top of Causey Pike to the top of Skiddaw — over the storm entirely — without wetting a feather. It was very interesting to watch its progress over the vale of Keswick, to see how it first obscured one place, then another, until a dark narrow mantle some miles in length, stretched entirely across it, from Barff to Keswick. After a little while, Bassenthwaite Lake could be discerned through the cloud, and soon the lake was cleared entirely, and it was easy for the eye to follow the progress of the storm in a south-east direction for several miles. We have only once before in our life seen the edges or boundary of a thunder storm so clearly defined; it was indeed a beautiful natural phenomenon. "Those who go down in ships to the great deep, see the wonders of the Lord." But the marvellous works of the Almighty and the manifestation of His power are more clearly developed to those who not only make the mountains the school for the study of natural objects, but their temple for worship. After this we continued our work industriously for a considerable time longer; now breaking up large stones, now hunting among the shingle as if we were seeking for a needle; sometimes lying down upon it for closer examination, and as our search extended over a considerable area, whatever we found that was worth taking, we left in "caché," with a mark to enable us to find it again. About four o'clock we began to collect them together, folding each fossil carefully in paper by itself, to prevent them rubbing and scratching each other, which would almost entirely spoil them; this work occupied about an hour. We then gathered up our washing, and returning to the sheepfold, we collected our "camp equipage." The weather being so unsettled, and as it was now so late in the season, we relinquished the principal object of our ramble for the next summer, reluctantly leaving the mountains, to return with a very heavy load to the Royal

Oak, at Braithwaite, according to our promise to the kind old landlord, who had begun to be uneasy on our account, as it had now been dark above an hour, and the night had become rather stormy. We found this short journey of three or four miles — from Causey Pike to Braithwaite — with our heavy load, was no easy task, and particularly as we took the rough mountain side direct for High Coldale, from which place there begins a road to the village of Braithwaite.

The next morning we set out for Barff, a distance of two miles, and as the roughest part of our work was over we intended to take the remainder easy; we therefore engaged a miner to carry our load all the way to that place, where there is a small road side inn, kept by Stephen Bowman, another homely and kind landlord; the landlady is from Furness, therefore sure to be kind. We had stopped at this place two nights in our former ramble, two or three weeks before, so we were received as an old acquaintance. We intended to stop a day here and hunt the immense shingle bed again, and in the morning take the coach to Cockermouth, and thence by rail to Furness in the evening.

When we came to the hill, which is close to the inn, we found a young gentleman there from Sheffield, working very industriously amongst the shingle. He told us he had been drawn thither by the writings of Professor Harkness. He was a member of a Naturalists' Field Club. It was very pleasant, he said, to hunt the coal measures for Organic Remains, as they were abundant in some places; he had never been amongst the old rocks before, and he thought he never would again. It was evident from his manner of working that he did not know how to look, nor what to look for, so we gave him a little lesson. We also showed him Professor Harkness's figures in the "Geological Quarterly Journal," but notwithstanding this, at midday neither of us had found a fossil. He now gave up entirely, and went back to Keswick, where he had been staying for a few days, we directed him to old Joseph Graham's little shop, where we thought he could buy a few fossils from the Skiddaw Slate. After the young gentleman left us we hunted with greater care and earnestness than before, and at length succeeded in finding a single specimen containing several entire figures of *phytograpsus* (Hall). This ended our day's work.

The next morning, to amuse ourselves until coach time, we went a fishing to Bassenthwaite Lake, which was close by; it was not for trout, pike, or char, there being no char either in Derwent or Bassenthwaite Lakes, but we assisted in fishing for railway waggons, which had been swallowed up by the flood and storm of the 27th of September.

The railway from Penrith to Cockermouth was in course of construction. It passes on the north side of Keswick, and close to the town, thence westward, between Bassenthwaite and Derwent Lakes. Near Barff it curves rather sharply to the north, and skirts (when at its ordinary level) the west margin of Bassenthwaite for about three miles; it then turns westward again and points direct for Cockermouth, where it forms a junction with the Cockermouth and Workington line, at the Cockermouth Station, which is at the west end of the town. This line of railway does not just now, at the time we write, skirt the borders of Bassenthwaite Lake, but it passes through it, there being a great many acres of water on the outside of the embankment. They are now "tipping" many hundred waggon loads of stone into the lake, to form the new railroad. This stone comes from a heavy rock-cutting, a mile from Barff. Notwithstanding the work for permanent way is composed principally of stone, the storm and flood have given it a sad ruffling, that will cause a considerable loss to the contractors. If Wordsworth had been living now, he would have said it was a judgment on them for having desecrated with the snort of the iron horse one of the loveliest spots on earth.

Half-past ten o'clock, and the coach from Keswick is now at the inn door; we carefully lift our load inside the coach, mount to the top, and we are off. The first three miles of our road is parallel with the lake, and also with the railway works, which at some places interrupted our progress a little. Perhaps they thought it of no consequence, as the railway will sweep this coach off the road altogether in a few months, also that from Keswick to Penrith; none will then remain in the Lake District but the one from Keswick over Dunmail Raise to Ambleside.

About three miles from Barff we leave Bassenthwaite Lake altogether, turning westward through Embleton Valley. The rock formation on the south side of this valley, dips S.S.E., and as this ridge is a continuation of the hill of Barff, where the dip is N.N.W., it is evident that there is a synclinal axis, between those two points, which synclinal cannot, however, be found from the top of a coach.

We are now through the whole of the Cumberland mountains, there being nothing north of Embleton but Lambfoot and Dunthwaite Fells, which are comparatively low. We continued on the Skiddaw Slate formation, westward, through Cockermouth, which forms a junction with the Mountain Limestone near Brigham. The whole of the stratifications of Coniston, Ireleth, &c., are wanting in this part of Cumberland.

At Cockermouth we have the junction of the rivers Derwent and Cocker; their united stream forms a noble river — now swollen by the heavy rains to an unusual height, filling, and in some places overflowing its banks. Cockermouth Castle is situate on a high and narrow spit of land at the confluence of those two rivers, and no military assault on either of these two sides could be successful, with the rivers in their present condition, for if either man or horse were to attempt to cross either of them, they would be swept away in a moment. It does not require much military capacity to see that the site of this castle is exceedingly well chosen; and more remains of the fortifications exist than of any border castle we know, that is not a regular garrison.

In contemplating these great accumulations of waters, that roll with resistless force on both sides of the castle, and which unite their streams at the foot of the hill on which it stands, the mind is naturally drawn to consider where it all comes from. Has the mound or barrier which impounds the waters of some great lake broken down, and let the waters run off as if the weir of a mill dam had burst and wasted them in some way? No, there are other reasons why the united streams of Derwent and Cocker should be greater in times of flood than any other rivers in Cumberland—but we will speak of this by and by; in the meantime it will be interesting to glance slightly at the river drainage of the whole of the Lake District.

Now, if a person is supposed to start from this point and to keep on the summit of drainage through the whole of the lakes, he would almost traverse a complete circle, which circle would include nearly all the central mountains of Cumberland. All the streams and rivulets on his right hand would be tributaries of the rivers Derwent and Cocker; those on the left would take various courses, which will be indicated below. To commence then in a northerly direction, the first streams on the left hand would be the head springs of the river Ellen, which rise near Torpenhow, takes a south-west direction by Outerside, Crosby, and Dearham, and fall into the Frith of Solway at Maryport. (Maryport was formerly known as Ellen Foot, and many country people call it by that name to this day.) The next is the river Waver, which rises on the Catland Fells, flows north by Waverton, and falls into the Frith of Solway in Abbey Holme. The Whampool rises near Rose Castle — the palace of the bishop of Carlisle — flows north-west by Wigton and falls into the Frith of Solway at Kirkbride. The Calder rises between Skiddaw and Saddleback, flows eastward between Carrock and Bowscale Fells, then turns due north by Hesket-Newmarket, Sebergham, and Dalston, and falls into the Eden, at Carlisle; from Carlisle the Eden flows north-west, about six miles, and falls into the Frith of Solway at Rockcliff. Not far from the source of the Calder, on Bowscale Fell, issue some springs of the river Greta, which flow by Threlkeld, into Derwent Lake. The Petterill rises near Graystock Park, flows eastward to Thornbarrow, then turns N.N.W. by High Hesket, Wray, and Carlton, and falls into the Eden about a mile east of Carlisle.

Our route is now south, and keeping to the summit of drainage, by Dowthwaite Head, Dod Fell, and Helvellyn, to near Raise Gap, the water running to the right hand, or westward, falls into the Vale of St. John, consequently into the Greta, and to Derwent Lake. The waters on the left into Patterdale, Ullswater Lake, and thence by the river Eamont by Broughton Castle, and into the Eden about four miles east of Penrith. The river Lowther rises on Barbeck Fells, twelve miles north of Kendal. We will now suppose we are there, and about to continue our journey from thence westward, and that we are still on the summit of drainage. The springs or runners on our right flow north; those on the left are the head springs of the river Mint, which flow south and fall into the river Kent, about a mile north of Kendal. Barbeck Fells is also the source of the river Lune. Four miles west on Hartop Fell the waters on our right fall into Haweswater Lake, thence by the Lowther, by Bampton, Askham, to join with the Eamont at Broughton Castle. Those on the left are the sources of the river Kent, which flow by Kendal and fall into Morecambe Bay, at its extreme north point, near Levens Hall.

After crossing the Pass of Nan-beild and High Street, the waters on our right flow into Ullswater Lake, and thence by the Eamont, the outlet of this lake to join the Eden as before stated. Those on the left are the spring heads of the river Troutbeck, which fall into Windermere Lake. We now cross the Kirkstone Pass, and Dow Crags. The waters on our left fall into the river Rothay, and thence into Windermere. We are now at Raise Gap again; the small runner on our left hand is the fountain head of the river Rothay; the water on our right falls into the Vale of Wythburn, and the Vale of St. John, and thence into Derwent Lake. Proceeding westward, and passing on the north side of Langdale Pikes, by the Stake and Cringle Crags, the waters on our right supply the falls of Lodore; those on the left are the sources of the river Brathay, and a little further west, on Wrynose, near the Shire Stones, the fountain head of the river Duddon, which runs south by Seathwaite and Ulpha, and falls into the sea at Hodbarrow, the south-east point of Cumberland.

At Bow Fell, and Eskdale Hawes, we have the fountain head of the river Esk, which flows south-west through Eskdale and falls into the sea at Ravenglass. At Sty-head Pass we have the head spring of Borrowdale Beck, which flows north through Borrowdale, by Boulderstone, and falls into Derwent Lake; this is the last stream that falls into Derwent Lake from these mountains. Near Sty-head on our left are the sources of the river Irt, which flow southwest by Wastdale Head, through Wastwater, thence by Santon and Carlton, and fall into the sea at Saltcoats, near Ravenglass. From Sty-head Pass we now follow the ridge of Great Gable, the waters on our right flow northwards into Buttermere Lake; those on the left, north-west, and are the fountain heads of the river Liza, which flows into Ennerdale Lake, thence by the river Ehen, to Ennerdale and Egremont, and falls into the sea near Beckermont.

After passing Gunterthwaite and Red Pike, we come to Floutern Tarn, near to which we leave the Green Slate and Porphyry, which we have traversed from Dowthwaite Fell, south, to Raise Gap, and from Berbeck Fells to this point in a north-west direction. We now come upon the Skiddaw Slate, which continues for about four miles, by Knockmurton and Blake Fell, soon after which the Skiddaw Slate forms a junction with the Mountain Limestone, at Lamplugh. We have now passed through the whole of the mountains in this part of Cumberland. The country westward to the sea is gently swelling land.

From near Lamplugh the coal measures continue until we strike the sea-shore at Harrington. From Floutern Tarn the waters on our right flow either into Crummock or Lowes Water Lakes, except a brook which rises on the west side of Blake Fell, which flows north-west by Dean and Clifton, and falls into the river Derwent at Ribton. Thus it will be seen that in this long traverse, from Raise Gap to Harrington, a distance of twenty-seven miles, we have been keeping to the summit of drainage nearly the whole way, which would be something like a person walking round the ridge of a circular building. During rain, all the water falling inside would represent the area of drainage of the rivers Derwent and Cocker, and all the water falling on the outside would represent the area of drainage of all the other rivers enumerated above.

As we have before stated, nearly the whole of the great mountains of Cumberland are inclosed within this area, and as the rain-fall of a district is influenced to a considerable extent by its mountains, we may reasonably expect

GEOLOGISING UNDER DIFFICULTIES.

more than an average quantity here; besides, there are other geological conditions which have a tendency to raise suddenly the rivers Derwent and Cocker — which it is not necessary to recite here; but there is one peculiarity with respect to the river Cocker, which we will notice in this place. Soon after we cross Sty-head Pass, the waters on our right fall into Buttermere. This lake has its outlet at the north end, by which it flows into Crummock Lake, the outlet of the latter being also at the north end—*i.e.*, by the River Cocker, which flows nearly due north, through the vale of Lorton to Cockermouth. So far we have a direct drainage to the north, but all the streams near Lowes Water from the west, the north, and the east, fall into that lake, which has its outlet at the *south* end, and after flowing by this outlet due south about a mile, falls into Crummock Lake also, and without having time to mix with the waters of Crummock, it is thrown out again by the river Cocker near the place where it entered, and then flows nearly due north. Thus a narrow ridge of hills, of which Mireside forms a part, presents the remarkable phenomenon of the waters on one side flowing south, and the same waters, on the other side, flowing north. We know of no other example of any water or river changing its course thus abruptly from south to north, as stated above.

We have been induced to give this slight sketch of the rivers and drainage of the Lake District, from the circumstance that local geography has not hitherto been taught in our schools; but this is now about to be remedied, as in one or two of the best schools in Ulverston, this subject will soon be made a branch of instruction. Pupils will then no longer be heard speaking with considerable knowledge of the great mountains and rivers in distant parts of the world, and knowing all about the Nile, the Ganges, the Mississippi, the St. Lawrence, &c., yet ignorant of our own beautiful rivers — the Duddon, the Crake, the Brathay, the Rothay, the Kent, and Leven, or even the stream which flows through the centre of the town of Ulverston.

After passing the junction of the Skiddaw Slate and the Mountain Limestone, near Brigham, there is nothing of particular geological interest in our progress through Workington and Whitehaven, until we come to the splendid deposit of New Red Sandstone of Saint Bees Head. Soon after leaving Saint Bees, in our journey southwards, we have evidence of the great elevation of the coast line; the different railway cuttings are principally through shore material, and on the north side of Seascale, there are dunes, or banks, forty or fifty feet high, composed almost entirely of sand, gravel, and other shore material. These accumulations are comparatively recent, certainly they are subsequent to the deposition of the New Red Sandstone. This elevation of the coast line continues, more or less, to the extreme south point of Cumberland.

Proceeding southwards from Seascale, we continue on the New Red Sandstone until we come on the Eskdale Granite, on the north side of Ravenglass. The granite of Eskdale is the greatest deposit of that material in the north of England, it continues southward to Bootle, where it forms a junction with the Skiddaw Slate of Black Combe. Black Combe is an outlier or island of Skiddaw Slate, entirely detached from the Skiddaw Slate of the central mountains of Cumberland. It continues southwards to Sylecroft, where it forms a junction with the Green Slate and Porphyry, and about half a mile east of Sylecroft at Limestone Hall, we cross a narrow vein of Mountain Limestone, probably not an outlier, but a continuation of the great Furness deposit of that material, which also includes Hodbarrow and Dunnerholme, therefore it is probable that the present generation may see extensive mines of iron ore wrought under the estuary of Duddon. After crossing this narrow vein of limestone, and near to Kirksanton, we should come upon the Coniston Limestone, if that formation be continued south-west from Beck Farm, but if it be not the case, then the Green Slate and Porphyry will form a junction with the Coniston Flag near Holborn Hill, in Millom. The Coniston Flag ranges through the whole of the east side of Millom, and forms a junction with the extreme south-west end of the Coniston Grit, at Foxfield Railway Station.

After crossing Angerton Moss, we enter on the splendid deposits of Lower and Upper Ireleth Slate, which continue through Kirkby, and form a junction with the great Furness deposit of Mountain Limestone near Dunnerholme.

Proceeding southward, after passing the rich iron ore mines of the Barrow Hæmatite Iron and Steel and Mining Co., at Park, and the equally rich mines of Messrs. Kennedy Brothers, at Askham and Ronhead, we come to the junction of the Mountain Limestone and the New Red Sandstone, in Hag Spring Wood, near Millwood. This is the highest stratification in Furness, continuing southward by Furness Abbey to Roose, beyond which no rock of any kind is seen on the surface to the extreme south point of Furness. Between Furness Abbey and Dalton we come upon the Mountain Limestone again, which continues eastward by Stainton, Urswick, and Scales to the shore of Morecambe Bay, at Aldingham. Between Dalton and Lindal the railway cutting gives a very interesting section of the rock, showing veins and other indications of iron ore. There is also a tunnel, about 600 yards in length, cut almost entirely through this rock. At Lindal we leave the train, shoulder our heavy load, and travel home to Swarthmoor.

GEOLOGISING UNDER FAVOURABLE CIRCUMSTANCES.

To be foiled in the attainment of any object on which the mind has been fixed involves a feeling not so much of disappointment as of humiliation—our self-esteem is wounded, indeed, we are beaten men. This feeling has haunted us more or less for the last five years, *i.e.*, ever since September, 1863, when we promised ourselves the pleasure of sleeping for a few nights at one of the numerous Rock Hotels, either on Causey Pike or Outerside. We proceeded there accordingly, and after spending two days and one night, we were reluctantly forced from the mountains by a storm of wind and rain such as is rarely witnessed in this country, and although the principal part of our object was defeated for that time, the project was not abandoned altogether, but laid by for a while to be taken up again at some future opportunity. On reading the highly valuable memoir on the Graptolites of the Skiddaw Slate, by Henry Alliyne Nicholson, D.Sc., M.B., F.G.S., we could not resist the temptation to visit the Cumberland mountains again, and to spend a couple of days at our old home, the Rock Hotel, on Outerside, near Causey Pike. This little journey was soon accomplished without hairbreadth escapes or adventures of any kind, we did not even fall in with a modern "Johnny Armstrong" with his wild band of moss-troopers to disturb our peace. It was quite a common-place every day excursion, all plain sailing and sunshine, suitable for "parlour geologists," or as our friend Mr. S. calls them, "the kid glove fraternity of the hammer and satchel." It was as follows:—On the morning we left home our better half packed in our wallet sufficient eatables to serve for two or three days, not forgetting our half-pint tin can which had done good service on former occasions as both coffee-kettle and coffee-cup, and although she could not prevail on us to promise we would not sleep on the mountains, we arranged that we should engage a young man at the village of Braithwaite to go with us and carry our burthen up to Causey Pike, and to show him our bedchamber, so that he would know where we might be found in the event of illness or accident which, however, was not very probable. In our former rambles on the mountains it was generally at the latter end of summer, in every instance in which we have patronised the Rock Hotel, it was late in September, but this was the end of May and the beginning of June, when we may reasonably expect beautiful weather, and it would be a luxury for any lover of nature to enjoy two or three nights on a sweet heather bed in a nice sheltered nook in the open air.

We left home by the early train for Whitehaven, thence by Workington and Cockermouth to Braithwaite, four miles west from Keswick. The village of Braithwaite is a central station for geologising in the Skiddaw Slate, several favourable localities for Organic Remains being within a few miles, viz., Barff, Mireside, Causey Pike, Outerside, &c. We put up at the Royal Oak, a small inn, where we were so kindly treated in September, 1863, by the then landlord, old Isaac Fearn. We secured the services of an active young man to accompany us and carry our load up the mountain. We started next morning in good time, and in less than two hours we were at our old lodgings at Outerside, near Causey Pike. We informed our companion, this would be our home for two days, and he must come for us the next evening at seven o'clock, and carry our load down to the Royal Oak again, when he would find us to a certainty, either dead or alive. While giving these instructions we were putting our provisions by in a suitable place. We then went forward two or three hundred yards to the shingle bed, accompanied by the young man, who, as he was anxious to see what we were going to seek amongst the stones on the hill side, remained with us about an hour, during which we found one or two very indifferent specimens of *carycaris*. When he left us he said, "I'se nert much i' love wi' yer wark amang t' rocks an' shingle o' Outerside Hill, but I'll come and fetch ye to-morrow neet at t' proper time." We wrought industriously all day, but did not succeed very well, finding we were following the line taken by Dr. Nicholson and Professor Harkness, and we even condescended to take two or three specimens which those gentlemen had either rejected or mislaid. When night came on, we left the shingle bed and returned to our home — the deserted sheepfold — pulled an armful of heather roots for our fire, and as the fine weather had dried up all the bog-water on the hill side, we had to go down the hill to the brook for water to make our coffee. We then lighted the fire, and after it had burnt down put the half-pint can (filled with water and coffee) on the hot embers, and our supper was ready. As we had not eaten anything since morning we did justice to it. After recommending ourselves to the Almighty we laid down on the bare dry ground against the wall of the sheepfold and slept till morning. We arose with the sun about four o'clock, and worked well for two or three hours, then went down to the brook for water, pulled more heather roots for our fire, made our coffee as before, and with a thankful heart enjoyed our breakfast. The morning was rather cold about sunrise, but it now became warm and pleasant; the view was delightful, and we sat for a while contemplating its wondrous beauty, but as the most glowing picture language of words is inadequate to do justice to the natural scenery by which we were surrounded, it would have been great presumption and a sort of physical impudence for us to attempt it. We could imagine the Genii of the mountains whispering in our ear a rebuke something like the following, "Now John, be quiet, you have said enough, and have got out of your depth already, you may flounder on from bad to worse. Don't meddle with subjects so far above your reach, gaze and admire for a while in silence, then pick up your old hammer and be off to the shingle bed, *that* is your place in nature." We took the hint and went accordingly, but we found that Dr. Nicholson and Professor Harkness had taken the cream off the greater part of it since our visit in 1863, and we only found one specimen of a branched or compound graptolite, which was folded and knotted in a singular manner. We then left the loose shingle, and wrenching off large pieces from the rock *in situ* we split them up, but we did not succeed much better by that method, and as it was past mid-day we thought we would have some dinner (a very unusual thing during work time). However, we would not take the trouble to make a fire, but carried the provisions down to the brook in the bottom of the deep valley. This brook which was so white and furious that it would have swept a horse away on September 27th, 1863, was now humble enough, the dry weather having made it very low in water, and a few small trout from two to four ounces each were seen, some of which we could have captured with our hands and made a very luxuriant dinner indeed. We hear some one ask, "will you eat your fish raw?" No, and we will give you the true recipe for cooking trout on the mountains, presuming you will not find it in M. Soyer or Mrs. Glass. *Catch* your trout first, then light your fire, and as it consumes renew it until you have sufficient of hot ashes. Take a piece of clay, temper it into a paste and flatten into a cake, next sprinkle a trout over with salt without opening it, wrap it up in clean paper, place it on the cake of clay which must be rolled round the fish so as to make it into a clay pudding. It is now ready for baking in the hot ashes; cover your pudding over, but be sure you do not make the clay red hot; take it out of the ashes in thirty or forty minutes, break the crust of clay and you have fresh fish for dinner; the edible part of the fish will leave the interior unbroken and undisturbed, and it may be removed in a lump, leaving the remainder perfectly clean.

Trout cooked as above may be eaten by the first ladies in the land, even the Princess of Wales, God bless her! if we should catch her on Causey Pike or Outerside would not reject the old man's dinner if cooked in this fashion. The late Professor Wilson would have been delighted with this sort of repast. We have no doubt he has often dined in a similar way, on the mountains of Cumberland, and if we had offered a couple of trout to Jean, a celebrated beauty of the Gordon family, she would have replied, "Na, na, my man, I will not take your fish, for this is no new sport to me, I will catch a bit trouty for myself." Her shoes and stockings would be off in a minute, her coaties kilted up a little, only just sufficient to keep her clothes dry, and show her splendid and graceful figure to advantage, then lightly stepping into the brook she would fish for her own dinner, for we have heard she was an expert hand at this mode of fishing when she went into the water in good earnest.

The above may be taken as our experience in cooking a fish dinner on the mountains, but on this particular day we dined in another fashion. Not taking the trouble to light a fire, we filled our little can with clear water from the brook, put coffee and sugar into it, and as we had no tea-spoon, we stirred it round and round with the shaft of our smaller hammer, which did very well. We cannot leave this sweet place without an attempt to describe our dining hall. Let the reader picture to himself an old man, bare-headed, seated on a rock by the side of the brook in the arena of an immense amphitheatre, shut in on the south by the towering Causey Pike, on the west by the mountain of Grassmoor, which presents its serrated and scarped western front to Crummock Lake, on the north-west by the mountain Whiteside, which overlooks all the northern part of Cumberland and the south of Scotland, with Scalehill, the vale of Lorton, and the river Cocker close at its foot; on the north by Coldale and Coldale Hawse; but, open on the east for a few miles, Bassenthwaite Lake reposing in the bright sunshine, under the western brow of the "Mighty Skiddaw" in the middle distance.

After we had finished our repast and returned thanks, we sat gazing for awhile on this enchanting scene until our eyes filled with tears, then we took up our hammer and wallet and ascended the hill for three or four hours' work, but on examining our store of fossils, we found we had obtained almost a sufficient wallet load, therefore we relaxed in our search for more, and would only take a few of the best, leaving those we rejected for a future visit. The day being clear and delightful, we sat down now and again to gaze and contemplate on the wonderful and sublime scenery by which we were surrounded, and thus in a dreamy mood the imagination would revel in all manner of impossible speculations. No signs of animal life could be seen, neither man nor beast, not even a single sheep in the whole range of view, and in all our geological rambles amongst this central group of mountains we had never seen a human being — we had a world to ourselves. Scenes like these we count as the principal pleasures of geological pursuit, for geology brought us there and the same science we hope will bring us there again. The hours seemed to fly quickly; evening drew on, and we repaired to the old sheepfold, and at seven o'clock our young man, true to appointment, arrived to carry our load down the mountain to the village of Braithwaite. When he spoke to us we felt somewhat ashamed at our effeminacy, and apologised by informing him we were bound by a promise to our wife to make this arrangement, and we must respect it and keep our word, moreover both she and our two sons experienced uneasiness on our account when we were here in September, 1863, and these mountains were then visited by extraordinary storms of wind and rain. Our family had arranged if we did not arrive at home the next night our elder son* would have set off in the morning direct to the inn at Elterwater, to gain some intelligence respecting our route up the rocky gorge between the mountains of Grassmoor and Whiteside, afterwards taking his course towards Causey Pike as the likeliest place to find us. He had set his mind on it, and nothing could restrain him. The young man shouldered our wallet, and we trudged down the mountain side together. Near the farm-house of High Coldale we passed several young men standing in the road, joking and bantering each other in a friendly manner. They spoke to our companion and eyed us rather knowingly, without comment, but they exchanged meaning glances with each other as much as to say, "that's t' auld crack'd feller that sleeps on t' mountains, and Willie ——— is bringing him down fra Causcy Pike, where he has bin for t' last two days, for I saw Willie and him ganging up togither." Our young man afterwards informed us the above was something like the remark one of them had made after we passed, but he said they were good lads for all that. A little while before we left off work we found a considerable portion of a trilobite. We know not if the genera can be determined, as it is wanting both head and tail, but there is sufficient to prove that it does not belong to the genus *Æglina*, of which we found a new species in September, 1863.

After a comfortable night at the Royal Oak, we left Braithwaite by the early train on the Keswick line for Penrith, thence to Carnforth and Ulverston.

* Alas, he is gone from us! He died in February, 1867, in the 25th year of his age, of an illness occasioned by severe study as a medical student, and we are left to lament him, for he was not only our son, but a dear and intelligent companion. Many delightful geological rambles we have had together when he was only a child.

FURTHER OBSERVATIONS ON THE HÆMATITE ORE OF FURNESS.

In a former part of this work we alluded to the immense masses of Hæmatite Ore deposited in various places within a few miles of Ulverston. We also stated that we possessed a great number of facts and specimens bearing on the question of the origin or formation of Hæmatite Iron Ore, and we crave the indulgence of our readers while we again allude to the above specimens to say they will be at the service of any gentleman who may be inclined to attempt the solution of this difficult problem. Various theories have been given to account for its origin, but all of them beset with difficulties. Its *entirely* igneous origin has been ably refuted by Mr. A. Moon, F.G.S., in the following manner:—" Some persons speak as if the hæmatite flowed in a molten condition into its present receptacles from some immense reservoir, located in the interior of the earth, but we see no reliable evidence to warrant us in entertaining such a theory, in fact the evidence points in a different direction. The presence of thin shale partings, interstratified with the iron ore; the laminated structure of the deposit in many places; the finding of shells occasionally in the beds of ore, and the absence of everything like the results of igneous action from the sides of the clefts through which this supposed molten metal must have passed, forbid our crediting such an hypothesis even for a moment. If this burning fluid had flowed into the fissures and cavernous spaces where we now find immense masses of hæmatite, it would have left some significant marks of its passage through the rocks. It would have calcined the clay; it would have materially altered the texture of the surrounding limestone; it would have hardened the shale. But no such indications are to be found in any part of the mineral field." Nevertheless, we must state that two or three of our specimens *seem* to have been subjected to heat. We also stated that we had noticed at the *open works* of the Barrow Hæmatite Iron and Steel Co., at Mousell, what *appeared* to be the effects of heat in several places on the face of the rock of ore, and we cannot entirely ignore this evidence, although it is the only fact we possess in favour of the igneous origin of Hæmatite Iron Ore, and even this may be fallacious.

A considerable portion of the present work has been in manuscript for eight or nine years, and at the time it was written only one fossil of perfect Hæmatite Ore had been obtained in the whole mining district of Furness, but within the last two years we have obtained upwards of two hundred fossils from the hæmatite mines of Lindal Moor. A great number of these are perfect iron ore, but there are specimens amongst them showing all the different gradations of change from perfect limestone to perfect iron ore. They are all, however, of marine origin and characteristic fossils of the Mountain Limestone. We also possess specimens of the phenomenon of metamorphism or change by imperceptible degrees from perfect granite to perfect iron ore (see p. 84). We offer no theory of our own, but certainly this fact points directly to that propounded by Mr. H. C. Sorby, F.G.S., of Sheffield, given as a quotation from Mr. A. Moon's "Geological Notes," (p. 66.) "As to the formation of the Cleveland Iron-stone, a theory was propounded some years ago by Mr. H. C. Sorby, F.G.S., to the effect that this iron-stone was an altered limestone rock, in which the carbonate of lime had been changed to carbonate of iron, by the percolation of water holding a salt of iron in solution. Mr. Sorby points to the shells frequently found in the iron-stone as evidences of his theory.

These shells have been changed to a certain depth, by the carbonate of lime being replaced by carbonate of iron; and in a similar manner he argues that the grains of the Oolitic limestone have been replaced by the carbonate of iron." In a letter to Mr. A. Moon (Feb. 6th, 1866,) Mr. Sorby observes,— " Since I wrote it " (meaning the article advocating the above theory) " I have prepared artificially many pseudomorphs of the carbonate of iron replacing carbonate of lime, and changed shells in a similar manner; but, since I was obliged to use proto-chloride of iron, I was compelled to use a higher temperature than would probable be sufficient with the carbonate." The satisfactory solution of the foregoing problem, viz., the ORIGIN of Hæmatite Ore is of the last importance to Furness, and should a discussion or paper war commence respecting it amongst our high geological authorities (which is not improbable) we may be able from our long experience, and from facts bearing on the subject, to furnish munitions for the fight which may ensue, quietly awaiting the result and looking on at a safe distance, but no temptation shall induce us to mingle in any way amongst the combatants.

The Hæmatite Ores of both Furness and Cleator are found almost entirely in large sops or pockets. There is no regular vein or lode known at present in either of these two centres of action. Perhaps the extensive excavations at Lindal Moor, and especially the mines of A. Brogden, Esq., M.P., at Stainton, may be considered the nearest approaching to a true vein. We are glad to see that the Barrow Hæmatite Iron and Steel and Mining Co. has commenced " boring operations " in earnest at Hawcoat, and if the work be persevered with until they have penetrated through the Permian rock, it is *probable* either coal or iron ore will be found below. This opinion is formed in part from our knowledge of the brown hæmatite mines of A. Brogden, & Co., of Mwyndy, near Llantrissant,

Glamorganshire. The ground was first broken at this place in a Calcaro-Magnesian Conglomerate. The brown vein "dipped" at an angle of 33°, and was wrought open until the vein was followed "but" up to the coal measures (about ninety yards), it then undershot the coal at the same angle for upwards of fifty feet, at which point the ore is lost in the floor of the incline. It was then followed down by a perpendicular shaft. Now when we consider that the principal vein of the brown Hæmatite Ore of Mwyndy *undersheds* the coal measures to an extent of fifty feet, and maintains its place, it naturally suggests the propriety of persevering with "deep borings" in the southern part of Low Furness, for there we have the coal measures, although we *may not* have the coal. To determine this point the ordinary method of proof is "deep boring," and we have no doubt the powerful and liberal Company which has undertaken the task will carry it through successfully. It may be objected that the brown Hæmatite Ore of Llantrissant is very different from the red Hæmatite of Furness. We admit there is a considerable difference between them both in colour and texture, but in chemical composition the difference is not very material, as may be seen from the following comparative analyses. The Furness ore contains more Silica, and the Llantrissant more lime, but the physical conformation of that portion of Glamorganshire is exceedingly like Low Furness, the same strata, the same order or sequence, viz., the Mountain Limestone, Magnesian Limestone, and the Coal Measures.

No Organic Remains have yet been discovered in the Llantrissant hæmatite, even our own careful seeking as well as that of others were unsuccessful. Mr. Brogden's superintendent of the works at Mwyndy (Mr. Vivian), has discovered in the ore some of the most splendid objects the mind can conceive. These consist of microscopic crystals of the most brilliant colours, none of which is intermixed with any other, each crystal standing clear in its own beautiful colour, and when a group of them is under the microscope, the eye can scarcely bear to look upon it. We have no doubt our Furness mines will furnish a rich feast for the microscopist, when he brings our own beautiful Hæmatite Ore under the instrument in earnest.

HOAD AND OUTRAKE.

There is no definite boundary between High and Low Furness,—it is almost entirely arbitrary, and we may approximate to it by a line drawn east and west from about Bouth to Broughton, yet the hill country extends southwards to the immediate neighbourhood of Ulverston, ending abruptly with the hills of Gamswell, Flan, Outrake, and Hoad. In the eastern portion of this tract from Ulverston to Bouth, Organic Remains are exceedingly scarce. We have never discovered a fossil on either Gamswell or Flan Hill, although we have visited both places several times. We have succeeded somewhat better at Hoad and Outrake, but they are very scarce there also. At the quarry at the south-east corner of Hoad, we obtained a very beautiful specimen of *Calymene brevicapitata*, and following westward the Hoad and Outrake range we have occasionally obtained a few specimens of three or four species of fossils, most of them corals. We had made several visits without obtaining Organic Remains, but it was no disappointment, as the view is worth a day's journey to any one who can enjoy delightful natural scenery, and this is of no ordinary kind. However we dare not attempt a description, it must be seen to be appreciated. One day after traversing the southern slope of Outrake without success, we tried the enclosure called Blundell's Plantation, situated immediately behind Flan How, the residence of J. Fell, Esq., J.P. Commencing with hammer and chisel on the first rock *in situ*, it broke with a splintery fracture, and after displaying a few small fossils, all of which were mutilated more or less, we developed a splendid group of trilobites all in a mass, but in attempting to detach them from the rock, the whole group broke up into fragments consisting principally of shapeless masses, but the heads were entire, the eyes of all being more perfect than any we had ever seen before, the eye-facets as clear and sharp as if they had life. Three or four of these are in our own collection, but more are in one of the cabinets of A. Brogden, Esq., M.P. We now visited this place frequently — on several occasions accompanied by our lamented geological friend, the late Rev. Francis Evans,* but we never found another trilobite, which was to be regretted, as we believe the whole are of new species, and during all these visits our patient friend never betrayed the least disappointment or vexation,—on the contrary his conversation was always pleasing, edifying, and instructive, and every way suitable for a Christian minister.

* Many of the pleasantest and happiest days of our life were spent in company with Mr. Evans and our friend Mr. W. Salmon, in examining the rocks of Furness and Cartmel. Some years ago, when Mr. Evans was in a delicate state of health, we induced him to join us in a weekly geological excursion : he soon became an enthusiastic lover of the science, and regained his health and strength. It would be out of place here to give a sketch of Mr. Evans, yet we cannot help putting on record our admiration of one whose society was so enjoyable. Nowhere did his rich and varied attainments, and the unaffected simplicity of his character, shine with a purer lustre than when, throwing off all professional reserve, he abandoned himself with the simple faith of a child to the study of Nature's works.

Pre-eminently he was a man who

"Looked through nature up to nature's God."

The recollection of those happy days is as refreshing to the feelings as the strains of a fine old melody which touch the inmost springs of a man's soul.

ANALYSES OF IRON ORES.

CLEATOR MOOR IRON ORE.

(By A. Dick.)

DESCRIPTION.—Compact red hæmatite; easily scratched by a file; lustre, earthy; colour, deep red-gray; streak, bright red; fracture, uneven, showing numerous cavities lined with microscopic crystals. This ore does not contain quartz visibly diffused through it.

Tabulated Results.—Ore dried above 100° C.

Peroxide of iron	90·36
Protoxide of manganese	0·10
Alumina	0·37
Lime	0·71
Magnesia	0·06
Phosphoric acid	trace.
Sulphuric acid	trace.
Bisulphide of iron	0·06
Insoluble residue	8·54
	100·20

Insoluble Residue.

Silica	7·05
Alumina	1·06
Peroxide of iron	0·19
Lime	trace.
	8·30
Iron, total amount	63·25

A trace of lead was detected in 500 grs. of ore.

GILLBROW ORE, ULVERSTON.

(By A. Dick.)

Description.—Red hæmatite; unctuous; easily scratched by the file; lustre, sub-metallic; colour, purplish red; streak, bright red; fracture, uneven and minutely crystalline; pieces of carbonate of lime and other minerals occur in it, which, getting coloured by the powder, connot be seen until the specimen is washed.

Tabulated Results.—*Ore dried above* 100° C.

Peroxide of iron	86·50
Protoxide of manganese	0·21
Lime	2·77
Magnesia	1·46
Carbonic acid	2·96
Phosphoric acid	trace.
Sulphuric acid	0·11
Insoluble residue	6·55
	100·56

Insoluble Residue.

Silica	6·18
Alumina, containing a trace of iron	0·30
	6·48
Iron, total amount	60·55

A whitish metal, precipitable by sulphuretted hydrogen from the hydrochloric acid solution, was found. The quantity obtained from 500 grs. of ore was so small that it could not be identified.

HÆMATITE, LINDALE MOOR, NEAR ULVERSTON.

(By J. Spiller.)

The sample was selected from a large quantity of the ore, consisting of fragments of various degrees of hardness, the majority of which were of the hard compact variety, deep grayish purple in colour, and covered with a brownish red unctuous powder; there were also small quantities of fibrous hæmatite and specular iron, together with quartz and a little earthy matter.

Tabulated Results.

Peroxide of iron	94·23--94·27
Protoxide of manganese	0·23
Alumina	0·51
Lime	0·05
Magnesia	trace.
Phosphoric acid	minute trace.
Sulphuric acid	0·09
Bisulphide of iron	0·03
Water, hygroscopic	0·39
,, combined	0·17
Insoluble residue	5·18
	100·88

Insoluble Residue.

Silica	4·90
Alumina	0·12
Peroxide of iron } Lime	traces.
	5·02
Iron, total amount	65·98

A distinct trace of arsenic was detected in 1680 grs. of ore.

FURNESS IRON ORE.—KIDNEY ORE, STAINTON.

(By Dr. Lyon Playfair.)

Peroxide of iron	97·93
Silica	2·03
Alumina	traces.
Manganese	,,
Carbonate of lime	,,
,, of magnesia	,,
Sulphate of lime	,,
Phosphate of lime	,,
Oxide of tin	0·03
	99·99

Lindal Cote Puddling Ore.

Peroxide of iron	90·76
Silica	5·43
Carbonate of lime	3·73
Tin	traces.
Copper	,,
Phosphoric acid	,,
Magnesia	,,
	99·92

Lindal Cote North Pit.

Peroxide of iron	90·93
Silica	6·68
Alumina	traces.
Manganese	0·24
Carbonate of lime	1·48
,, of manganese	traces.
Sulphate of lime	0·40
Phosphate of lime	traces.
Oxide of tin	0·02
	99·75

Eure Pits Dark Ore.

*Peroxide of iron	64·15
,, of manganese	15·35
Titanic acid	4·00
Silica	16·50
	100·000

* Equal to 44·88 metallic iron, and 2·4 metallic titanium.

Note.—Mr. Cameron says this is a very good sample of dark ore, and ought to be very valuable in steel making.

THE LLANTRISSANT HÆMATITE IRON ORES.

(By Dr. H. M'Noad.)

No. 1.—Yellow Ore. (Large.)

Water	9·400
Peroxide of iron	75·540
Oxide of manganese	none.
Carbonate of lime	11·070
Phosphoric acid	0·127
Insoluble residue	3·200
	99·337

Iron % 52·88.
Iron % in the roasted ore 58·4.

No. 2.—Yellow Ore. (Small.)

	1st Sample.	2nd Sample.
Water	10·40	8·40
Peroxide of iron	87·13	68·60
Oxide of manganese	none.	none.
Carbonate of lime	0·90	0·90
Phosphoric acid	0·13	0·15
Insoluble Residue	1·60	21·60
	100·16	99·65

1st Sample.—Iron % 61·12.
    Iron % in roasted ore 67·8.
2nd Sample.—Iron % 48·0.
    Iron % in roasted ore 52·3.

No. 3.—Grey Ore. (Small.)

Water	9·20
Peroxide of iron	80·00
Oxide of manganese	none.
Carbonate of lime	1·00
Phosphoric acid	0·12
Insoluble residue	10·00
	100·32

Iron % 56·0.
Iron % in the roasted ore 62·0.

No. 4.—Grey Ore. (Large.)

Water	8·00
Peroxide of iron	77·60
Oxide of Manganese	none.
Carbonate of lime	0·90
Phosphoric acid	0·15
Insoluble residue	14·00
	100·65

Iron % 54·3.
Iron % in the roasted ore 59·0.

No. 5.—Blue Rock Ore.

	1st Sample.	2nd Sample.
Peroxide of iron	68·50	74·00
Oxide of Manganese	none.	none.
Carbonate of lime	traces only.	1·90
Insoluble residue	32·00	23·60
Phosphoric acid	·18	·18
	100·68	99·68

1st Sample.—Iron % 48·00.
2nd Sample.—Iron % 51.76.

Dark Brown Ore.

	No. 6. Small.	No. 7. Large.
Water	8·00	6·000
Peroxide of iron	89·71	91·430
Oxide of manganese	3·20	0·280
Carbonate of lime	traces.	traces.
Phosphoric acid	0·12	0.125
Insoluble residue	2·80	2·200
	100·83	100·035

Small Ore.—Iron % 62·8.
    Iron % in roasted ore 68·0.
Large Ore.—Iron % 64·0.
    Iron % in roasted ore 68·0.

A qualitative analysis having shown that the "Insoluble Residues," of all the samples had the same composition, the residues of all the nine samples were mixed together, and a quantitative analysis made of the mixture.

Composition of the Insoluble Residues:—

Silica	93·5
Alumina, with a little Peroxide of iron	6·0
Lime	-traces.
	99·5

Dr. M'Noad observes: "A striking and valuable feature in these ores is their freedom from sulphur and the insignificant quantity of *phosphorus* which they contain. They are exceedingly rich and valuable ores, especially the '*dark brown;*' and *when roasted, would serve admirably for steel-making by Bessemer's process*, for which they are especially adapted, in consequence of the minute quantity of *phosphoric* acid which they contain. I consider these ores to be very valuable, either for working alone, or with Welsh Mine."

CLEVELAND ORE.

(By A. Dick.)

DESCRIPTION.—Chiefly a carbonate of protoxide of iron; lustre, earthy; colour, greenish gray; streak, similar; fracture, uneven, showing here and there small cavities, some of which are filled with carbonate of lime. Throughout the ore are diffused irregularly a multitude of small oolitic concretions, together with small pieces of an earthy substance resembling the ore but lighter in colour. When a mass of this ore is digested in hydrochloric acid till all carbonates and soluble silicates are dissolved, there remains a residue having the form of the original mass of ore. It is extremely light, and falls to powder unless very carefully handled. It contains the oolitic concretions, or else skeletons of them, which dissolve completely in dilute caustic potash, showing them to be silica in a soluble state. Under the microscope some of them are seen to have a central nucleus of dark colour and irregular shape, but none of them present any indication of organic structure or radiated crystallization. If the residue, after having been digested in caustic potash, be washed by decantation, there remains a small number of microscopic crystals; some of these, which are white, are quartz, and others, which are black and acutely pyramidal, consist chiefly of titanic acid. Professor Miller, of Cambridge, succeeded in measuring some of the angles of the crystals containing titanic acid, and found that they correspond to similar angles in anatase. The green colour of the ore seems to be due to a silicate containing peroxide and protoxide of iron, but this could not be exactly determined because it was not found possible to dissolve out the carbonates without acting at the same time upon the silicate of iron.

Tabulated Results.—*Ore dried at* $100°$ C.

Protoxide of iron	39·92
Peroxide of iron	3·60
Protoxide of manganese	0·95
Alumina	7·86
Lime	7·44
Magnesia	3·82
Potash	0·27
Carbonic acid	22·85
Phosphoric acid	1·86
Silica, soluble in hydrochloric acid	7·12
Sulphuric acid	trace.
Bisulphide of iron	0·11
Water, in combination	2·97
Organic matter	trace.
Insoluble residue (of which 0·98 is soluble in dilute caustic potash), and consists chiefly of oolitic concretions	1·64
	100·41
Iron, total amount	33·62

Insoluble Residue.

Silica	1·50
Alumina, with a trace of peroxide of iron	0·10
Titanic acid, about	0·03
Lime	trace.
	1·63

No metal precipitable by sulphuretted hydrogen from the hydrochloric acid solution of about 1200 grs. of ore was detected.

GLEASTON.

WHILE this work was passing through the press our friend, Mr. W. T. Swainson, of Newland, has drawn our attention to a deposit at Gleaston, in which he has discovered many rare and interesting Organic Remains, either in actual contact, or so near, as to be greatly affected by the Trap Dyke so ably described by E. W. Binney, Esq., F.R.S., F.G.S., to the Manchester Literary and Philosophical Society. We also have made a slight examination, along with a young geological student, and examined several beautiful specimens in the possession of S. Salt, Esq., but have a diffidence in giving any opinion. We hope, however, this notice will attract the attention of geologists to this new field for enquiry, and that a satisfactory opinion may be arrived at respecting its nature and contents.

APPENDIX.

LIST OF FOSSILS

Derived from localities in Cumberland, Westmorland, and parts of Lancashire and Yorkshire, drawn up by Professor M'Coy.

CAMBRIAN FOSSILS FROM THE SKIDDAW SLATE TO THE CONISTON FLAG INCLUSIVE.

Graptolites latus (M'Coy).
,, sagittarius (Hisinger, sp.)
,, ludensis (Murchison).
Palæopera interstincta (Wahlemberg, sp.)
,, var. subtubulata (M'Coy).
,, megastoma (M'Coy).
,, petalliformis (Lonsdale, sp.)
,, tubulata (Lonsdale, sp.)
Favosites crassa (M'Coy).
Nebulipora explanata (M'Coy).
,, papillata (M'Coy).
Stenopora fibrosa (Goldfuss, sp.)
,, var. a, Lycopodites (Say).
Habysites catenulatus (Linnæus).
Sarcinula organum (Linnæus).
Petraia æquisulcata (M'Coy).
Berenicea heterogyra (M'Coy).
Plitodictya explanata (M'Coy).
Retepora Hysingeri (M'Coy).
Caryocystites Davisii (M'Coy).
Tentaculites annulatus (Schlotheim).
Beyrichia strangulata (Salter).
Lichas subpropinqua (M'Coy).

Ceranrus clavifrons (Dalman, sp.)
Zethus atractopyge (M'Coy).
Odontochile obtusi-caudata (Salter).
Portlockia apiculata (Salter).
Chasmops Edini (Eichwald, sp.)
Calymene brevicapitata (Portlock).
,, sub-diademata (M'Coy).
Homalonotus bisulcatus (Salter).
Stotelus Powisii (Murchison).
Illemus Rosenbergii (Eichwald).
Spirifera biforata (Schlotheim).
,, var. b. dentata (Pander).
,, var. d. fissicostata (M'Coy).
,, insularis (Eichwald).
,, per-crassa (M'Coy).
Pentamerus lens (Sowerby, sp.)
Orthis actoniæ (Sowerby).
,, calligramma (Dalman).
,, crispa (M'Coy).
,, expansa (Sowerby).
,, flabillatum (Sowerby).
,, parva (Pander).
,, plicata (Sowerby, sp.)
,, porcata (M'Coy).
,, protensa (Sowerby).
,, vespertilio (Sowerby).
Leptœna deltoidea (Conrad).
,, var. b. undata (M'Coy).
,, minima (Sowerby).
,, sericea (Sowerby).
,, transversalis (Dalman).
Strophonema antiqua (Sowerby, sp.)
,, grandis (Sowerby, sp.)
,, pecter (Linnæus, sp.)
,, spiriferoides (M'Coy).
Leptagonia depressa (Dalman).
Lingula Davisii (M'Coy).
,, ovata (M'Coy).
Pterinea termistriata (M'Coy).
Cardiola interrupta (Broderip).
Ortheceras filosum (Sowerby).
,, laqueatum (Hall).
,, vagans (Salter).
,, subundulatum (Portlock).
,, tenuicinctum (Portlock).
Cycloceras annulatum (Sowerby, sp.)

Cycloceras ibex (Sowerby).
,, subannulatum (Munster, sp.)
Lituites cornuarietis (Sowerby).

SILURIAN FOSSILS FROM CONISTON GRIT TO UPPER LUDLOW
ROCK INCLUSIVE.

Nebulipora papillata (M'Coy).
Stenopera fibrosa (Goldfuss, sp.)
,, var. b. regularis (M'Coy).
Halysites catenulatus (Linnæus, sp.)
Cyrthaxonia Siluriensis (M'Coy).
Spongarium æquistriatum (M'Coy).
,, interlineatum (M'Coy.)
,, interruptum (M'Coy.)
Actinocrinus pulcher (Salter).
Taxocrinus orbigni (M'Coy).
Tethyocrinus pyriformis (Phillips, sp.)
Uraster primævus (Forbes).
,, Ruthvenii (Forbes).
,, hirudo (Forbes).
Protaster Sedgwickii (Forbes).
Tetragonis Danbyii (M'Coy).
Cornulites serpularis (Schlotheim).
Tentaculites tenuis (Sowerby).
Serpulites dispar (Salter).
Trachyderma squamosa (Phillips).
Beyrichia Klodenii (M'Coy).
Ceratiocaris elliptica (M'Coy).
,, inornata (M'Coy).
,, solenoides (M'Coy).
Odontochile caudata (Brongmart, sp.)
,, var. minor
Calemene tuberculosa (Salter).
Homolonotus Knightii (Konig).
Forbesia latifrons (M'Coy).
Euryterus cephalaspis (Salter).
Liphonotreta Anglica (Morris).
Discina rugata (Sowerby).
,, strita (Sowerby).
Spirifera subspuria (D'Orbigny).
Spirigerina reticularis (Linnæus, sp.)
Hemythynis navicula (Sowerby).
,, nucula (Sowerby, sp.)

Orthis lunata (Sowerby).
Strophomena filosa (Sowerby, sp.)
Chonetes lata (V. Buch. sp.)
Lingula cornea (Sowerby)
Avicula Danbyii (M'Coy).
Pterinea Boydii (Conrad, sp.)
,, dimissa (Conrad, sp.)
,, lineata (Goldfuss).
,, pleuroptera (Conrad).
,, retroflexa (Wahlemberg, sp.)
,, var. naviformis
,, subfalcata (Conrad, sp.)
,, tenuistriata (M'Coy).
Cardiola interrupta (Broderip).
Modiolopsis complanata (Sowerby, sp.)
,, solenoides (Sowerby, sp.)
Anodontopsis augustifrons (M'Coy).
,, bulla (M'Coy).
,, securiformis (M'Coy).
Orthonotus semisulcatus (Sowerby, sp.)
Sanguinolites anguliferus (M'Coy).
,, decipiens (M'Coy).
Leptodomus amygdalinus (Sowerby, sp.)
,, globulosus (M'Coy).
,, truncatus (M'Coy).
,. undatus (Sowerby, sp.)
Grammysia cingulata (Hisinger, sp.)
,, var. b. triangulata (Salter).
,, var. g. obliqua (M'Coy).
,, extrasulcata (Salter, sp.)
,, rotundata (Salter).
Arca Edmondii formis (M'Coy).
,, primitiva (Phillips).
Cucurrella coaretata (Phillips).
,, ovata (Sowerby, sp.)
Nucula Anglica (D'Orbigny).
Tellinites affiinis (M'Coy).
Conularia cancellata (Sandberger).
,, sublilis (Salter).
Pleurotomaria crenulata (M'Coy).
Murchisonia torquata (M'Coy).
Naticopsis glaucinoides (Sowerby, sp.)
Hollopella cancellata (Sowerby, sp.)
,, gregaria (Sowerby, sp.)
,, intermedia (M'Coy).
Litorina corallii (Sowerby, sp.)

Litorina octavia (D'Orbigny, sp.)
Bellerophon expansus (Sowerby).
Orthoceras angulatum (Wahlemberg).
,,   baculiforme (Salter).
,,   bullatum (Sowerby).
,,   dimidiatum (Sowerby).
,,   imbricatum (Wahlemberg).
,,   laqueatum (Hall).
,,   subundulatum (Portlock).
.,   tenuicinctum (Portlock).
Cycloceras ibex (Sowerby).
,,   subannulatum (Munster, sp.)
,,   tenui annulatum (M'Coy).
,,   tracheale (Sowerby).
Hortolus ibex (Sowerby, sp.)

NOTE.—The letters "sp." (species,) where written after a name signify that the author quoted gave the name of the species, but with a different generic term.

LIST OF FOSSILS FOUND BY THE AUTHOR IN THE LOCALITIES SPECIFIED.

B. Birkrigg; G. Gleaston; H.F. Hawkfield; L. Lindal; and S. Stainton.

RETEPORA. *This genus may be found in several of our limestone quarries, but especially at Hawkfield.*

Retepora irregularis, H. F.
,,   undulata, H. F.
,,   polyporata, H. F.
,,   laxa, H. F.

MILLEPORA. *This genus occurs in more than one of our limestone quarries.*

Millepora interporosa, H. F.
,,   specularis, H. F.

Millepora occulata, H. F.
,,   rhombifera, H. F.
Calamopora tumida.
,,   megastoma.

ZOOPHYTA.

Syringopora ramulosa, B.
Cyathophyllum regium, B.
Poteriocrinus Egertonii, and several other species of encrinites at Hawkfield.
Modiola granulosa, G.
Pleurorhynchus armatus, L.
Pinna costata, H. F.

Brachiopoda.	Spirifera.
Producta Martinii, B.	Spirifera senilis, G.
,, costata, B.	,, crenistria, G.
,, antiquata, B.	,, septosa, G.
,, comoïdes, B.	,, squamosa, G.
,, edelburgensis, B.	,, resupinata, G.
,, latissima, B.	,, arachnoïdea, G.
,, aurita, B.	,, papillionacea.
,, quincuncialis, B.	Euomphalus bifrons, B.
,, scabricula, B.	,, cristatus, L.
,, muricata, B.	Pleurotomaria interstrialis.
,, concinna, B.	,, atomaria, B.
,, lobata, B.	,, sculpta, B.
,, cetosa, B.	Bellerophon hiulcus, L.
,, analoga, B.	,, urii, L.
,, depressa, B.	Goniatiles Listeri, S.
,, gigantea, B.	,, bidorsalis, S.
,, pectinoides, B.	Orthoceras undulatum.
,, mesoloba, B.	,, inequiseptum.
,, punctata, B.	
,, fimbriata, B.	Trilobites.
,, granulosa, B.	
,, spinulosa, B.	Asophus granuliferus, G.
,, pustulosa, B.	,, seminiferus, G.
,, rugata, B.	

Note.—The above list of Organic Remains does not contain all the species we have obtained in Furness. Several specimens, both Silurian and Carboniferous, particularly three or four beautiful species of trilobites from the "Scrow" at Coniston, are now in our National Museums, and we have no doubt many new species of carboniferous fossils may be obtained both at Hawkfield and Gleaston Castle.

ELEVATION AND SUBSIDENCE OF THE SEA COAST OF FURNESS.

It is a well-known geological fact that there have been considerable elevations and subsidences of the dry-land, extending, in some instances, over wide areas; and we have not to go to remote regions — to Sweden or the American continent, for proofs of this fact.

There is evidence of both these phenomena in Furness, and we will give an example of each, with a few remarks on the physical conditions of the district at the time of their occurrence.

First, then, with reference to subsidence of the land, we would advise any one who is desirous to examine and judge for himself, to take a quiet walk southwards from the village of Biggar, in the Isle of Walney, at or about low-water of a twenty feet tide, and, at the distance of about a mile, he will come to a forest of timber trees, broken off about half a yard above the ground, all of them *in situ*, having their roots firmly fixed, and they can be clearly traced through all their ramifications amongst the rocks and stones of the scar. It is evident that when this forest grew and flourished, even the roots of those trees were above the reach of the highest tides, and as this mass of tree-stumps can only be seen at low-water of a twenty feet tide, which, by the tide table, rises thirty-six feet at this place, it is very clear there has been a subsidence or sinking of the land of at least thirty-six feet.

From the above evidence we have a right to assume that such has been the fact, and if so, what would be the physical condition of this district before the subsidence commenced? Something like the following: — Walney and Piel joined to the main-land of Furness and Millom would be extended southward some miles into the Irish Sea, and there would be dry land in that direction wherever there are now only six fathoms of water; there would be no Estuary of the Duddon, and the mouth of that river would be far below Hodbarrow Point. Moreover there would be neither Ulverston nor Lancaster Sands, and except for the rivers Leven, Kent, and Kear, a person would be able to travel on dry land from Roose Beck by

Cartmel Wharves and Humphrey Head to Hest Bank, so that Morecambe Bay would be very small indeed.

Now, although we admit there has been a subsidence on this part of the coast to the extent recited above, there is also evidence that there has at one time been a greater sinking than exists at present. The proof of this is in the fact that between Roose Beck and Rampside there is a shell bed several feet in thickness, not only in the sea-cliff, several feet above high-water mark, but also in the enclosed land, a considerable distance from the sea shore. This immense bed of sea shells is composed all of existing species, principally the *Cardium Edule*, or common cockle, also the panfish, pecten, papshell, &c., such as abound on the coast at this moment. One portion of the bed is a conglomerate, cemented together with carbonate of lime, but it is principally a mass of loose shells. There can be no doubt this shell bed was not deposited all at one time, but by slow degrees. It is also certain that the lowest part of the bed was deposited first, and as the land subsided, more and more shells were laid above the first — stratum above stratum, until the whole were deposited. Even the top of the shell bed does not determine the limit of *this* subsidence of the land, for we have evidence of a greater occurrence at another place, and as the highest tide at the present time does not reach the top of the bed by several feet, it is clear that a process of elevation has since taken place equal to the difference between ordinary high-water mark and the top of the shell bed. Supposing this difference to be fifteen feet, what would be the physical condition of this district if it were now at the point of greatest subsidence? A considerable portion of the low land between Roose Beck and Rampside would be flooded by every eighteen feet tide; Roosecote Marsh would be almost permanently covered; there would be two small Islands of Walney instead of one; the low and fertile lands between Conishead Priory and Ulverston, as well as the whole of the moss lands from Ulverston to Greenodd, would be overflowed by every tide. Plumpton Woods and Hoath would be islands, and vessels of 100 tons burthen would be enabled to load and discharge their cargoes on the lawn in front of Birk Dault House; Winder Moor and Cartmel Moss would be overflowed; Old Park, Low Frith, and Roudsea Woods would be little islands, and a high tide would almost reach the Duke of Devonshire's door in Holker Park; Haverthwaite and Carke would be sea ports, and Gleaston a fishing village for boats of ten or twenty tons burthen; Ulverston would be a principal sea port, the tide flowing to Sunderland Terrace and the foot of Quay Street.

This may appear wild, but it is nevertheless a true picture of what would be the physical aspect of the district under consideration, if an ordinary spring-tide reached the top of the highest shell bed at Roose Beck. With reference to the lapse of time since this subsidence and elevation of the land took place, although it has been probably thousands of years, (for there is no exact geological chronology) the fact of the trees in the submerged forest of Walney being of exogenous type, or increasing by annular rings of growth, and the shells in the Roose Beck deposit of existing species, — as a geological epoch, it is very recent. Indeed it is only as yesterday compared to the time when the Mountain Limestone of Birkrigg, Baycliff, and Stainton were deposited as a calcarious mud at the bottom of the sea, and of the immensely more remote period when the Silurian rocks of Hoad and Kirkby moor, and the yet more ancient Cambrian rocks of Coniston and Millom were deposited under the like circumstances.

Although we have no geological evidence of the fact, we may reasonably suppose that during the deposition of the Roose Beck shell bed, Furness was not inhabited by man. We may also infer that the towns of Liverpool,

Manchester, and Preston, had no existence, therefore no inhabitants to devour our delicious shell fish, and from the immense quantity in the deposit alluded to above, we may believe that the sands of Morecambe Bay, and the Estuary of the Duddon were almost paved with those dainties, which would live and die undisturbed except by gulls and other sea fowl, swarming along the sea-coast of Furness.

This may not have been the case entirely, however, (for it is only a fanciful speculation,) yet the evidence of elevation and subsidence of the sea coast of Furness remains perfect, which was all we attempted to prove.

EARTHQUAKE AT RAMPSIDE, BARROW, &c.

Now that the excitement occasioned by the earthquake of the 15th of February, 1865, at Rampside and Barrow, has subsided, and people can speak about it with calmness, many circumstances connected with it have come to our knowledge which have not been reported before, several of them from our own observation, and others from reliable sources. As it was our intention from the first to ascertain, if possible, how far its effects had been felt in every direction, and whether violent or otherwise, with a view of laying down and colouring the same on the ordnance map, showing the different gradations of intensity, by shading; it was, therefore, necessary to inspect, personally, every place which had suffered injury from its effects.

For this purpose we proceeded to Rampside, calling at Roose Cote, and Moorhead. At Roose Cote, they were commencing to repair Mr. Ross's chimneystack; Messrs. T. and W. Huddleston's chimneys were both somewhat injured. In passing Moorhead, we noticed some large pieces of stone lying on the ground, blackened with soot, and upon inquiry the mistress of the house said they were part of their chimneys, which had fallen outside, but far more of them fell into the house, and two of her children had a narrow escape. One, a little girl, was sitting by the fire, with an infant on her knee, the mother, providentially, being close by, had just time to snatch them away from almost certain destruction. This good woman took us through her house, and showed us that the walls in every room were cracked, and some of the flags in the floor were broken by the effects of the shock. At the next house, (a cottage, adjoining the barn, the end wall of which was badly cracked,) the cottager was close by setting some roots of rhubarb. We asked if he felt the earthquake. "Feel it!" he exclaimed, "I was working at this midden when it com, and it ree roo'd ma about, just as if I woz in a riddle; I cud hardly stand."

After leaving Moorhead we passed Rampside Church, the Parsonage, and two other good houses close by, none of which appeared to be injured. These are all situated on high ground above the village, and with these exceptions, we believe, every house in Rampside was more or less injured by the earthquake, some of them to a fearful extent. The Post Office was very much shaken, and rendered very unsafe, as was also the residence of J. Clegg, Esq., a beautiful "cottage *ornée*," only one story in height. This house was then shored up with timber, notwithstanding which it was highly dangerous, and a considerable part of it had to be taken down at the time. But the most remarkable case of injury was that of a house close to high water mark, on the shore of Morecambe Bay. This dwelling is built of Red Sandstone, dressed, bedded, and jointed, but since its erection the front wall seems to have been whitewashed several times, which had almost entirely hidden the jointing. The earthquake not only made several large cracks in different parts, but almost every stone in the front showed a clear joint broken through the coat of whitewash; thus proving, that nearly every individual stone in the front of the house had been moved on its bed by the effect of the shock. It had to be taken down immediately. There were three or four other houses which it was deemed prudent to pull down also, as they showed a greater amount of damage after the breaking up of the frost. We called on Mr. Clegg, iron ore merchant, who kindly volunteered to go with us to see the effects produced by the earthquake on the railway near Conkle, and on the sands on both sides of it. On our road we met William Parker Simpson, landlord of Conkle Inn and John Thompson, of Roose Beck, fisherman, the two men who witnessed the extraordinary commotion on the sands,— the throwing up of sand, water, and stones. They are both respectable and creditable men, whose word may be depended upon. We were glad to meet with them, and they both cheerfully accompanied Mr. Clegg and ourselves to the sands on the west side of the railway, where there had been a crack in the earth thirty yards in length. It was then partly obliterated by carting, being in the road along the bottom of the railway embankment, but between the railway and West-field Point there still remained several hollows, or basins, from which at the time of the convulsions, and for some hours after, copious springs of water issued. We were anxious to hear all the particulars the men could give, as they were the only persons on the sands at the time of the occurrence.

W. P. Simpson made the following statement:—"John Thompson and I were coming from Fowla Island, one mile from Rampside, and when we had got nearly half way we saw at a distance from us, a great mass of sand, water, and stone, thrown up into the air higher than a man's head. It was nearly in a straight line between us and Rampside, and when we got to the place there were two or three holes in the sand, large enough to bury a horse and cart, and in several places near them, the sand was so soft and puddly that they would have mired any one if he had gone on to them. We thought this very strange, but we supposed it was owing to the frost, for we did not feel the least shock, or know anything of an earthquake until we got to Rampside, and saw that everybody was in terror, and the houses sadly shattered. We then went to Conkle, and found a crack in the ground at the foot of the railway embankment, about thirty yards in length, and water was boiling up in a great many places, just like the great spring at Bien Well. ["Bien Well," a copious spring of very pure water, on the shore of Morecambe Bay, five miles north of Rampside, yielding about 500 gallons per minute.] There were more than 300 of them, and they extended above half a mile on the sands towards Barrow, and at one place there were a great many in a straight line, and only two or three yards from each other."

In returning to Barrow, we followed the line pointed out by Parker Simpson, which is about north-west and south-east, and we saw the remains of several of the springs, but no water was flowing from them, as it was nine days after the occurrence of the earthquake.

From the foregoing account, and also from our own observations, it appears that the earthquake travelled from south-east to north-west, and the line of its greatest intensity was near the north shore of Barrow Channel, from Conkle to Westfield Point, and from a careful survey, we believe its effects did not extend over an area of more than seven square miles. The violence, the very limited area over which it extended, and other circumstances peculiar to this occurrence, place it in opposition to all the theories propounded as the producing cause of earthquake phenomena, especially that of Professor Rogers, of the United States of America, which is the most recent, and as it has been favourably received by geologists in general, it will be interesting to compare the Professor's theory with the effects produced by the earthquake at Rampside and Barrow, by which it will be seen that the cause assigned by him for producing earthquake agency, could not possibly have effected a violent disturbance of the surface of the earth over an area so limited.

The theory of Professor Rogers is as follows:—He considers the producing cause of earthquakes as an actual pulsation in the fluid matter beneath the crust, propagated in the manner of great waves of translation, from enormous ruptures, caused by tension of elastic matter, and floating forwards on its surface under the superimposed rocky crust of the earth. He also assigns " the thickness of the solid crust of the earth, to be twenty miles;" then, from mathematical reasoning, it is evident that no combination of forces, exerted at a depth of twenty miles, could cause a violent disturbance of the surface of the earth, over an area of about seven square miles, without affecting the district for many miles in every direction. Therefore, we may conclude the centre of disturbance which produced this earthquake, was not twenty miles below the surface of the earth, and it is highly probable it was not a tenth part of that depth.

In reviewing all the circumstances connected with this strange occurrence, there are two deserving especial notice. First, we have positive proof that the earthquake wave travelled from south-east to north-west, *i.e.*, from Rampside to Barrow, for the two men—William Parker Simpson and John Thompson—were on the sands at the time of its occurrence, in a south-east direction from Rampside, and only about two hundred yards from the eruption, yet they did not feel the least shock, or know anything of an earthquake until they were informed by the people of Rampside. This, we think, is a proof that the earth was not affected to a distance of 200 yards, in a south-east direction from the site of the principal eruption, and, if such were the fact, it almost demonstrates that the disturbing power was at no great depth below the surface of the earth.

The other circumstance is the fact of the great number of copious springs of water—upwards of 300—which burst up simultaneously on the sands between Conkle and Westfield Point. We are not aware that similar outbursts of water,—in the form of springs—have accompanied earthquake phenomena in any part of the world. Several learned men, among whom are, Kircher, Des Certes, Dr. Priestley, Dr. Stukely, Beccaria, Mr. Mallet, and Professor Rogers, have arrogantly propounded theories of their own, to account for earthquake phenomena, all differing essentially from each other, except those of Dr. Stukely and Beccaria; therefore, they cannot be all right, and it is highly probable they are all wrong. Why should we not, rather than speculate on matters beyond our reach, which, do not admit of proof, acknowledge with all humility, our entire ignorance on this mysterious agency, believing it to be a chapter in natural philosophy, the true interpretation of which, has not yet been revealed to man.

The earthquake was felt in Barrow with very different degrees of intensity, in various parts of the town. That portion of it, farthest from the harbour, on comparatively high ground, was only slightly affected; but the Strand, and other streets in the lower part of the town, were visited with considerable violence. Of this, we had personal experience, having been nearly thrown from our chair whilst writing, and had we been standing at the time it is likely we should have been thrown down. The shock, or rather movement of the earth, did not cease instantly, but seemed to continue for four or five seconds, for we had time to turn towards the window, and by noticing other buildings in the neighbourhood became convinced that there undoubtedly was a movement amongst them. Directly it was over, there occurred a scene of a rather ludicrous description, and although an earthquake is no laughing matter, we could scarcely refrain from smiling at what came under our observation. There is, in our neighbourhood, a sailmaker's large workroom, in the roof or top story of a high block of buildings, and the earthquake had only subsided a few seconds, when we saw all the sailmakers rush out of the bottom door like a swarm of bees, almost tumbling over each other to escape from the premises. They all ran across the street, and stared up at the top of the building, expecting to see the whole block come down, for the timbers in the roof of the sailroom had been in violent commotion, a sufficient cause for alarm in anyone. However, the sailmakers would not believe that the disturbance in the roof of their workshop was caused by an earthquake, and, as they could not see sufficiently from the street, they crossed over to a piece of vacant building ground, to enable them to have a better sight of the roof; waiting a considerable time before they again ventured to ascend to their place of business.

GLOSSARY OF TERMS.*

ACICULAR, needle-shaped (Lat. *acus*, a needle).

Actynolite (Gr. *actin*, a ray, and *lithos*, a stone), a frequent mineral in granitic compounds.

Alluvium, Alluvial tracts.

Amber (Arab. *ambar*), a fossil tertiary resin.

Amorphous (Gr. *a*, without, and *morphe*, regular form).

Amygdaloid, a frequent variety of trap rock.

Anthracite (Gr. *anthrax*, charcoal), a non-bituminous variety of coal.

Anticline, anticlinal axis.

Apatite, phosphate of lime or phosphorite, found among granitic rocks.

Asterophyllites, a coal-measure and Permian plant.

Astræa (Gr. *astron*, a star), silurian star-coral.

Atolls, the name given to coral islands of an annular form, that is, consisting of a circular belt or ring of coral, with an enclosed lagoon.

Augite (Gr. *auge*, lustre), the principal mineral in many trap and volcanic rocks.

Auriferous (Lat. *aurum*, gold, and *fero*, to yield), applied to veins and deposits yielding gold.

Avalanche (Fr. *avalanges, lavanches*), an accumulation of ice, or of snow and ice, which descends from precipitous mountains, like the Alps, into the valleys beneath.

Azoic (Gr. *a*, without; *zoe*, life), as applied to stratified rocks.

BASALT, as distinguished from green-stone.

Basin, trough, or syncline of stratified rocks.

Berg-mahl (Swedish) mountain-meal, a recent infusorial earth.

Bitumen (Gr. *pitus*, pitch of the pine tree), as a mineral compound.

Botryoidal (Gr. *botrys*, a cluster of grapes, and *eidos*, form), applied to certain concretionary forms.

Breccia and brecciated, from the Italian, composed of irregular angular fragments.

Bunter (Ger., variegated), a member of the trias.

Burrstone, a siliceous rock of the Paris tertiaries, used for millstones.

CAINOZOIC (Gr. *kainos*, recent; *zoe*, life), as applied to fossiliferous strata.

Calcairie grossier (Fr. coarse limestone), one of the eocene beds of Paris.

Calcareous springs, (Lat. *calx, calcis*, lime).

Carboniferous system (Lat. *carbo*, coal, and *fero*, to yield).

Cataclysm, (Gr. *kataklysmos*, an inundation), applied to any violent flood deluge.

Catenipora (Lat. *catena*, a chain, and *porus*, a pore or passage), a silurian coral.

Chalybeate springs (Gr. *chalybs*, iron), or those impregnated with iron.

* Page's Handbook of Geology.

GLOSSARY.

Chert, a peculiar flinty admixture, occurring in many limestones.

Chiastolite slate or schist, a rock of the clay-slate group.

Chlorite (Gr. *chloros*, greenish); chlorite schist, &c.

Chondrites, a species of seaweed.

Clay-stone, an earthy variety of felspar-rock; clay stone-porphyry.

Clinkstone, or phonolite (Gr. *phonos*, sound), a trappean rock.

Cornstones, limestones of the Devonian system.

Cosmogony (Gr. *kosmos*, the world, and *gone*, origin), reasonings or speculations respecting the origin of the universe.

Crag (Celt. *creggan*, a shell), a shelly tertiary deposit, found chiefly in Norfolk and Suffolk.

Crater (Gr. *krater*, a cup or bowl), the term applied to the cup-like orifice of volcanoes.

Crinoidea (Gr. *krinon*, a lily, and *eidos*, form), lily-like radiata.

Crystal (Gr. *crystallos*, ice), originally applied to transparent gems, but now used to denote all minerals possessing regular geometrical forms.

Cyathophyllum (Gr. *cyathos*, a cup, and *phyllum*, leaf), fossil cup-coral.

Cycloid, cycloidians, &c., a division of fishes.

DEBRIS, a convenient term adopted from the French for all heterogeneous accumulations of waste material.

Degradation (Lat. *de*, down, and *gradus*, a step), the act of wearing or wasting down gradually or step by step.

Denudation (Lat. *de*, down, and *nudus*, naked). The removal of superficial matter so as to lay bare the inferior strata, is an act of denudation; so also the removal by water of any formation or part of formation.

Detritus (Lat. *de*, from, and *tritus*, rubbed), matter worn or rubbed off rocks by aqueous or glacial action.

Dyke, wall-like masses of igneous matter filling fissures in stratified rocks are so termed.

EOCENE, or lower tertiary group.

Erosion (Lat. *erosus*, gnawed or worn away).

Euomphalus (Gr. *eu*, well, and *omphalos*, the naval), a coiled nautiloid shell of the mountain limestone.

Exuviæ (Lat. *exure*, to cast or throw off). In Zoology this term is applied to the moulted or cast-off coverings of animals, such as the skin of the snake, the crust of the crab, &c.; but in Geology it has a wider sense, and is applied to all fossil animal remains of whatever description.

FAULT, fissure or dislocation of strata.

Favosites (Lat. *favns*, a honey-comb), a silurian coral.

Felspar (Ger. *fels*, rock, and *spath*, spar), as a rock.

Ferruginous (Lat. *ferrnm*, iron), impregnated with iron; *ferriferous*, yielding iron.

Fibrous texture, composed of fibres like asbestos.

Fissile structure (Lat. *fissus*, capable of being split).

Flint, as a rock, formation of, in chalk system.

Foraminifera (Lat. *foramen*, an opening), a class of minute chambered shells, with an orifice in the septa or plates which separate the chambers.

Fossils, fossil remains (*fossus*, dug up).

Fuci, fucoids, sea-weeds, or fucus-like impressions in silurian rocks

Fuller's earth, a variety of absorbent clay, used in the scouring or fulling of woollen cloth.

GANOID, ganoidians, &c., a division of fishes.

Garnets, in metamorphic rocks, &c.

Gault (*provincial*), a member of the chalk system.

Glacier (Lat. *glacies*, ice), the term applied to those masses of ice which accumulate in the higher gorges and valleys of snow-covered mountains.

Glossopteris, a fern of the coal measures.

Graphite (Gr. *grapho*, I write), so called from its use in making writing-pencils. This substance consists almost entirely of pure carbon with a small per-centage of iron, the proportions being about 90 to 9. It is also termed *plumbago* and *black-lead*, from its appearance, though lead does not at all enter into its composition.

Graptolites, characteristic silurian zoophytes.

Greensand, a member of the chalk system.

Gryphæa (Gr. *gryps*, a griffin), a beak-like shell of the oolite.

Gypsum (Gr. *gypsos*, from *ge*, earth, and *epso*, to boil), originally applied to all limestones.

HELIOLITES (Gr. *helios*, the sun, and *lithos*), silurian corals.

Holoptychius, a fish of the upper old red and lower carboniferous ages.

ICEBERG (Ger. *eis*, ice, and *berg*, a mountain), the name given to the mountainous masses of ice often found floating in the arctic and antarctic seas.

Indurated, hardened by heat; and in this sense should be kept distinct from 'hard' or 'compact.'

JOINTS, divisional planes, "backs and cutters."

KAOLIN, a Chinese term for a fine pottery-clay derived from the decomposition of granitic or felspathic rocks.

LACUSTRINE or lake deposits.

Laminated (Lat.), composed of thin plates or laminæ.

Lava, an Italian term, now universally applied to all molten rock-matter discharged from volcanoes.

Lepidodendron, a carboniferous fossil.

Lignite (Lat. *lignum*, wood), a variety of coal.

Littoral (Lat. *littus*, the shore), applied to all deposits and operations taking place near or along the shore, in contradistinction to pelagic (*pelagus*, the deep sea) or deep-sea deposits.

MAGNESIAN limestone, a member of the Permian system. Any limestone containing a notable percentage of carbonate of magnesia is termed "magnesian."

Mesozoic (Gr. *mesos*, the middle, and *zoe*, life), as applied to fossiliferous strata.

Millstone grit, a subdivision of the carboniferous rocks.

Miocene, or middle tertiary group.

Moraine, a Swiss term for the mounds of detritds (sand, gravel, and boulders) brought down by glaciers.

Mountain limestone, or carboniferous limestone.

NEOCOMIAN, a synonyme of the greensand.

Neozoic (Gr. *neos*, new, and *zoe*, life), as applied to fossiliferous strata.

Neuropteris, a fern of the coal-measures.

Nummulite, a fossil of the lower tertiary.

OBSIDIAN (Gr. *opsianus*), a compact vitreous lava, or volcanic glass; so called from being polished by the ancients, and used for looking-glasses.

Ochre, hydrated oxide of iron, as derived from coal-measures.

Ornith-ichnites, fossil footprints of birds.

Orthoceras, Orthoceratite (Gr. *orthos*, straight, and *keras*, a horn), a genus of straight horn-shaped chambered shells.

Outcrop, or extreme edge of inclined strata.

Overlying, as applied to overflows of igneous rocks.

Palæozoic (Gr. *palaios*, ancient, and *zoe*, life), as applied to certain fossiliferous strata.

Pecopteris, a fern-like fossil in coal-measures.

Pelagic (Gr. *pelagos*, the sea), applied to deep-sea deposits and operations, as distinguished from shore or littoral ones.

Pentacrinite (Gr. *pente*, five), a five-sided encrinite.

Petrify, Petrification (Lat. *petra*, a stone, and *fio*, I become). All vegetable or animal matters found in a fossil state, or converted into stony matter, are said to be petrified.

Petroleum (Lat. *petra*, rock, and *oleum*, oil), literally rock-oil; a liquid mineral pitch, so called because it seems to ooze out of the rock like oil.

Pitchstone, and pitchstone-porphry, varieties of igneous rock so termed from their pitch-like lustre.

Placoid, Placoidians, &c., a division of fishes.

Pleistocene, a subdivision of the "Drift."

Pliocene, the upper group of the tertiary system.

Polype (Gr. *polys*, many, and *pous*, a foot), the zoological term applied to zoophytes having many tentacula or feet-like organs of prehension; hence also the term *polypidum*.

Porphyry (Gr. *porphyreos*, purple), originally applied to a reddish igneous rock used in Egyptian architecture, but now applied to all igneous rocks having detached crystals (mostly of felspar) disseminated through the mass; hence the term *porphyritic*. We have thus porphyritic granites, greenstone porphyries, felspar porphyries, trachytic porphyries, and so forth.

Portland stone, a stratum of the upper oolite.

Post-tertiary, or recent accumulations.

Potstone, a soft magnesian rock.

Pumice (Ital. *pomice*, allied to *spuma*, froth or scum), a light, porous, froth-like lava.

Pyrites, (Gr. *pyr*, fire), sulphurets of iron or of copper.

Quaquaversal (Lat. on every side) This term is applied to strata which dip in every direction from a common point or centre of elevation.

Ragstone, applied to coarse concretionary or breccio-concretionary rocks, as coral rag, Kentish rag, &c.

Retepora (Lat. *rete*, a net, and *porus*, a pore), a flustracea-like zoophote found in various formations.

Saccharoid (*saccharum*, sugar, *eidos*, like), like loaf-sugar in texture.

Saliferous (Lat. *sal*, salt, and *fero*, I yield), a term applied to salt-yielding strata.

Scoriæ (Ital. *scoria*, dross), volcanic cinders or cindery-like accumulations.

Section (Lat. *sectus*, cut), the line, actual or ideal, which cuts through any portion of the earth's crust so as to show the internal structure of that portion (just as one would slice a loaf, or saw up a tree), is termed a section.

Sediment (Lat. *sedere*, to settle down), various kinds, as river, lake, and oceanic.

Septaria (Lat. *septum*, a division or fence), applied to nodules of ironstone, &c., occurring in the shales of the coal-measures, lias, and other strata, because, when broken up, the interior is often divided into net-like compartments by minute veins of carbonate of lime.

Siliceous springs (Lat. *silex*, flint), or those holding siliceous matter in solution.

Silt, fine mud, clay, or sand deposited as sediment from water.

Stalactite and Stalagmite (Gr. *stalagma*, a drop).

Steatite (Gr. *stear*, fat), so called from its greasy or soapy feel; soap-stone.

Stigmaria, supposed root of sigillaria, and characteristic of carboniferous rocks.

Stratified rocks, synonymous with aqueous, and sedimentary.

Strike, the linear direction of any stratum as it appears at the surface.

Sub (Lat. under), often applied in geology to express a less degree of any quality, as sub-columnar, not distinctly columnar; sub-crystalline, indistinctly crystalline; applied also to position, as sub-cretaceous, under the chalk.

TALUS, the sloping accumulation of debris which takes place at the base of a cliff or precipice exposed to the weathering effects of frosts, rains, and other atmospheric agents.

Thermal (Gr. *therme*, heat), applied to hot-springs and other waters whose temperature exceeds 60 deg. Fahr.

Travertine, a limestone of modern formation.

Trilobites, characteristic silurian crustacea.

Tuff, or tufa (Ital. *tufo;* Gr. *tophos*), originally applied to a light porous lava or pumice, but now applied to all porous rocks; hence trap-tuff, calc-tuff, or calcareous tufa.

Unstratified rocks, synonymous with igneous and volcanic.

Volcanoes (Lat. *Vulcanus*, the god of fire), active, dormant, and extinct.

WACHE, (Ger.), a term applied to all soft earthy varieties of trap, whether tufaceous or amygdaloidal.

Warp, a local term for marine silt.

Whinstone, whin, a Scottish or Saxon designation for greenstone, but by miners applied to almost every hard or indurated rock that comes in their way.

GAZETTEER

which also includes six-figure National Grid References for localities and four-figure references for areas, lakes, etc. See inside cover of Ordnance Survey Maps for full explanation.
(........) identifies location;
[........] is name in vogue

	N.G.R.
ABBOTS HALL 166	395756
,, Holme 216	177507
,, Wood 35, 89, 95	220723
Aldingham vii, 10, 128, 221	283711
Allithwaite 178	387765
Allonby 166	0743
Ambleside 180, 214	3704
America 99	—
Angelton Farm 27, 66	213843
,, Moss 70, 99, 221	2285
Applethwaite Common 26, 59, 68	4202
Appletreeworth 51, 52, 54, 62	2492
Arnside 174, 177	4578
Ashes, The 13, 14	322798
Ashghyll Quarry 52, 54, 62, 63, 69, 76	270954
Askham [Askam] 33, 221	215777
,, Wood 76	220771
BACKBARROW 15, 16, 108	356850
Bala 98	—
Bampton 217	5118
Bankend (Duddon) 25	199884
,, (Grizebeck) 27, 29	233853
Bank Field 126	268746
,, House 72	234810
Banks Gill Beck 72	240814
Bailiff Ground 28, 72	228804
Barbeck Fells 216, 218	
Bardsea vii, 12	301745
,, Mill viii	299740
Barff 182-184, 211-214, 223	214267
Barrow 37, 253, 256-258	2070
,, Channel 256	2267
,, Waterworks 29	242782
Bassenthwaite Lake 182, 207, 211, 213, 214, 227	2229
Baycliff viii, 11, 81, 92, 107, 111, 113, 114, 252	288723
,, Hags 85	2872
Beacon Tarn 68	274900
Beckermont [Beckermet] 218	0206
Beck Farm 17, 25, 50, 88, 180, 221	165810
Beckside (Ireleth) 29	235823
,, (Scales) 128	263726
,, (Ulverston) 93	282802
Belle Isle 179	3996
Benson Knott 25	5494
Berwick-on-Tweed 113	—
Betsy Crag 19	306023
Bien Well [Bean] 10, 13, 256	290720
,, ,, Scar vii, 93	295719
Biggar 42, 250	192662
Bigland Scar 174	350825
Billing Cote 89	225726
Birk Dault 16, 252	344834
Birkrigg viii, 12, 81, 85, 92, 107, 110, 252	2874
Black Combe 87, 220	135855
,, Hall 19, 24	239012
,, Scars 14, 93	327770
Blake Fell 194, 218	111197
Bleansley Bank 26, 51, 97, 98	205890
Blease [Blisco] Pike of 20	272042
Blundells Plantation [Outrake] 234	288797
Bolton Farm 128	259729
,, Heads 34	255732
Bootle 96, 220	107882
Borderiggs 27	222873
Borrowdale 23,217	2614
,, Beck 217	2312
Boulderstone 217	254164
Bouth 233	329856
Bow Fell 217	2406
Bowscale Fell 216	1690
Braithwaite 183,196, 198, 203, 206, 208, 212, 222, 223, 227, 228	231237
,, Beck 207	2423
,, Brow 183	230240
Brandreth 19	276027
Brathay 17, 19, 26, 61, 76, 181	3602
,, Bridge 18, 43, 48	366034
,, Hall 18	366031
,, quarry 64, 65	3501
,, river 18, 19, 25, 91, 217, 219	3503
Brestmill Beck 95	216726
Bridge End 19	301029
Brigham 214, 220	275238
Brighouse 28, 72	228798

GEOLOGICAL FRAGMENTS

Brotherdale Head [Bretherdale] 59	575051
Broughton 26, 27, 61, 95, 98, 99, 233	2187
,, Castle 216, 217	213880
,, High Common 52, 62	2594
,, Mills 18, 32, 49, 51, 54, 62	223907
,, Tower Park 62	2188
Burney 68	2685
,, Farm 70	256858
Burton-in-Kendal 136	530765
Buttermere 195, 219	1717
,, Lake 218	1815
Butts Beck 33	234746
Calder river 216	0304
Canada 98	——
Capeshead 13, 173	333778
Cark 170, 252	363765
Carkettle 34, 77, 82, 83	248775
Carlisle 216	——
Carlton (Carlisle) 216	4355
Carlton 217	0899
Carnforth 228	
Carrock Fell 216	3433
Cartmel 174, 176, 177	379788
,, Moss [White] 252	3480
,, Parish 165	3878
,, Wharves [Wharf] 251	3668
Castlehead 175, 176	422800
Castle Rock of St. John 182	330188
Catland Fells 216	2441
Causeway End 27, 70	232862
Causey Pike 183, 196, 198, 201, 209-212, 222, 223, 226, 228	218209
Cautly-spout 68	683973
Channel House 81	264780
Chapel Island 14, 93	321758
Chappels 27	237839
Church Scar 93	292712
Cinderstone Beck 24	201920
Cleator 149, 231	0113
Cleator Moor 235	0315
Clerk's Beck 118-122	269744
Cleveland 230, 241	230241
Clifton 218	5326
Cocker, river 197, 214, 215, 218, 219, 226	1426
Cockermouth 182, 187, 212-214, 219, 223	1229
Cockley Bank [Beck] 24, 25	247017
Coldale 197, 198, 206, 207, 227	2122
,, Hawes 198, 227	1821
,, High 206, 212, 228	227227
Cold Pike 20	264035
Coldstream 112	——
Cold Well 17, 64, 76	355066
Colwith Bridge 19	330030
,, Force 19	329031
Commonwood Quarry 24, 44	204946
Conishead Bank vii, 12	310753
,, Priory 252	304758
Coniston 27, 45, 46, 54, 55, 58, 60, 61, 63, 64, 91, 95, 99, 252	332976
,, Lake 67, 68	3094
,, Monk 43	318984
,, Old Man 45, 53, 58	272978
,, Railway Station 61, 63	300976
Conkle 255, 256, 258	234660
Coupren Point 170	346743
Coup Scar 93	311753
Crake, river 91, 219	2987
Crier of Claife 179	385981
Crinkle Crags 20, 22, 23, 217	250050
Crosby 215	0838
Cross-a-moor 80	268767
Cross Gates 33, 82, 151-153	235752
Crooklands 34, 155	238745
Crosslands 95	213713
Crummock Water 183, 195, 218, 219, 226	1518
Cumberland 10, 19, 27, 48, 98, 99, 181, 187, 194, 196, 214, 215, 218, 220, 226	
Cunsey 17	382935
DALE-HEAD 24	241006
Dalston 216	3750
Dalton 34, 100, 138, 154, 221	230740
Dean 218	075253
Dearham 215	0736
Dendron 92, 93, 100	247707
Dent 60, 136	074253
Derwent, lake 182, 207, 213, 216, 217	2521
,, , river 182, 197, 214-216, 218, 219	2514
Doodles Quarry [Dowdales] 154	225745
Dove Bank 27	235844
,, Ford 27	235847
Dow Crags (Coniston) 45, 185	262978
,, (Kirkstone) 217	
Dowthwaite Head 216	3720
,, Fell 218	3621
Dragley Beck 80	293777
Duddon, Bridge 25, 26, 43, 48, 51, 54, 60-62, 97, 98, 108	197883
,, Estuary 10, 28, 37, 61, 87-89, 99, 107, 220, 250, 253	2080
,, Hall 24	194895
,, river 9, 19, 20, 23, 24, 27, 30, 31, 91, 217, 219	2090
Dunmail Raise 181, 214	327117
Dunnerdale 43, 45	2394
,, Hall 24	214954
,, Fells 24	230940
Dunnerholme 28, 31, 74, 76, 88, 220, 221	211798
Dunthwaite 214	174328
EAGLE CRAG (Brathay) 18	
,, ,, (Thirlmere) 182	
Eamont, river 216, 217	4827
Eccleriggs 66	211865
Eden, river 216, 217	5144
,, vale 180	5144
Egremont 218	0111
Egypt 85	
Ehen, river 218	0915
Elbow Scar 93	283703
Ellen river 215	1238
Ellenfoot 215	0437
Elliscales 33	225747
Elleray 68	4199
Elterwater 9, 43, 45, 228	3304
Elwood Scar 14, 93	320753

GAZETTEER

Embleton Valley 214	1429	HAG SPRING WOOD	
Ennerdale 218	1015	(Dalton) 34, 35, 89, 221	2173
Esk, river 217	1802	Hag Spring Wood	
Eskdale 23, 41, 96, 193, 217, 220	1802	(Plumpton) 80	311787
,, Hawes 217	2308	Hallbeck 89	228700
Esthwaite 17, 68, 91	3697	,, , Old 91, 92	233695
Eure Pits 238	240755	Hammerside Hill 93	314774
Europe 99	—	Hampsfell 178	3978
		Hard-knot 24	232025
FAIR VIEW 81	283782	Harlock 69, 163	253804
Fell Side 81	267785	Harrington 218	9926
Ferry—Windermere 71, 179	3995	Hartley Ground 26, 51	215898
Fingal's Cave 83	—	Hartop Fell [Hartsop] 216	4111
Flan Hill 82, 85, 100, 233	283797	Haverigg 88	160787
Floutern Tarn 218	1217	Haverslack Hill 30, 33, 34	2378
Flodden Field 112	—	Haverthwaite 252	341838
Flushes Wood [Ashes] 80	323798	Hawcote [Hawcoat] 10, 36, 42, 85, 87, 94, 95, 100, 231	203720
Ford (Berwick) 112	—	Haweswater 216	4814
Foul Bridge 27	235849	Hawkfield 113, 124, 126, 127, 128	260732
Fowla [Foulney] Island 255	247638	Hawthwaite 62	217892
Foxfield 27, 61, 65, 66, 68, 70, 99, 221	209854	Hawkshead 64	352982
Frith 174	338798	,, Hill 63	335989
,, , Low 13	340797	Hazlehurst Point 13, 174	335800
Frizington 187	0317	Head Crag 28, 66, 70	229827
,, Moor 186	0420	Heathwaite Fell 70	2586
Furness Abbey 10, 35, 42, 87, 89, 94, 95, 97, 100, 137, 221	217717	Hell-Ghyll 20, 21 23	259055
		Helvellyn 216	3415
,, , High 15, 86, 165, 233	—	Hensingham 193	9926
,, , Low 10, 36, 41, 49, 84, 86, 91, 109, 114, 165, 232, 233	—	Hesket 180, 216	4344
		,, High 216	4745
		,, Newmarket 216	340386
		Hest Bank 251	—
GALE NASE CRAG 18, 25	3702	Hestham 88	151800
Gamswell 82, 85, 233	274796	High Cockan 95	190710
Ganges 116, 219	—	,, Cross 26, 99	207877
Gargrave 72	230807	Highfield House 34	245740
Gawthwaite 68, 69, 100, 158, 159	268848	High Haume 31-34, 71, 74, 76, 77, 82	226762
German Ocean 112, 113	—	,, Kinmont 84, 96	123907
Gillbanks 81, 100, 142, 143	283790	,, Mere Beck 28, 29, 72	228795
Gillbrow 34, 80, 236	255763	,, Park 19	323029
Gillhead 68	243825	,, Stott Park 16	375885
Gillwood 72	236812	,, Street 217	4411
Gleaston 35, 92, 98, 100, 115, 243, 252	255708	Hill, Millom 50	180830
,, Castle 115, 128-130	262715	Hindpool 37	195700
,, Park 35	265704	Hinning House 24	240000
Godderside Marsh 172	3477	Hoad 71, 81, 85, 100, 105, 179, 233, 252	295791
Goldrill Dub 24	234989	Hoath [Oath] 252	316796
Grange 165, 174, 175, 177, 178	4077	Hodbarrow 76, 89, 217, 220	183783
Grasmere 180, 181	3306	,, Point 30, 31, 88, 89, 250	182781
Grassmoor 183, 196, 197, 226, 228	175204	Holborn Hill 88, 221	171804
Graystone House 25, 26, 50	187875	Holker 174	361770
Graythwaite 16	372909	,, Hall 172	359774
Great Gable 218	2110	,, Park 13, 14, 80, 171, 252	3577
Greenburn 45	290023	Holling-house 24	230970
Greenodd 15, 252	315826	Holme Island 175, 178	422782
Greenroad 61	190840	Holmes Green 77	233761
Greenscow Wood 76	219759	Holy Island 113	—
Greta, river 181, 216	2924	Hopkin Ground 62	227908
Grey Friars 45	260004	Hougholm Point 170	385740
Greystock Park [Greystoke] 216	4132	Humphrey Head 166, 169, 170 176, 251	3973
Grizebeck 70, 100	238850	Hunter How Coppice 19	—
Grizedale Head 68	334951		
Groffacragg Scars 69, 100, 158, 159	268837	IDRIDGE SCAR 93	305715
		Ingleborough 156	—
Gunterthwaite 218	—	Ireleth 30, 31, 67, 72	224776
		Irish Sea 10	—

268 GEOLOGICAL FRAGMENTS

Iron Yeats 69, 100, 158, 159	267818
Irt, river 217	1303
Isle of Man 156	—
,, of Wight 96	—
JEFFY KNOTTS WOOD 19	3503
KEAR, river 250	5573
Kendal 130, 216, 217	5292
Kent, river 216, 217, 220, 250	5194
Kentmere 26, 59, 180	456041
,, Tarn 68	456020
Kents Bank 165, 166	400760
Keswick 182-185, 194, 195, 197, 198, 211, 213, 214, 223, 228	2623
Kiln Bank 24	213940
Kirkbride 216	—
Kirkby 72, 95, 100, 158, 221	2282
,, Hall 27	235835
,, Ireleth 70, 85	2282
,, Mill 72	236823
,, Moor 28, 29, 67, 69, 70, 75, 76, 82, 85, 100, 158, 163, 252	2582
,, Park 70	236865
,, Quarries 75	250838
Kirkflat Tarn 122	267742
Kirkhead Cavern 166	393755
,, Hill 166, 167, 168	393757
Kirkhouse 24	197934
Kirksanton 87, 88, 220	143803
Kirk Scar 10	292713
Kirkstone Pass 217	4008
Knockmurton Fell 218	095191
Knotthollow 69, 100, 156-159, 163-165	270802
LAMBCRAGS [Longcrags] 45, 58	317004
Lambfoot Fell 214	1530
Lamplugh 193, 194, 218	089209
,, Cross 185, 194	076200
,, Hall 194	089207
Lancashire 19	—
Lancaster Sands 250	—
Lane End 62	219900
Langdale 23	2905
,, , Great 20	3006
,, , Little 19	3103
,, Pikes 22, 217	2707
Latterbarrow 68	367991
Leatherbarrow 45	
Leece 92	244695
Leonard Scar 93	270687
,, Hill 93, 115	267687
Leven, river 9. 10, 15, 18, 20, 91, 181, 220, 250	3483
,, , viaduct 14, 108	3278
Levens Hall 217	495851
Levy Beck 114	2777
,, ,, Bridge 109	286773
Lickle, river/valley 26, 51, 61, 62, 98, 99	2190
Lightburne Park 81	290778
Limestone Hall 31, 76, 87, 89, 220	138813
Lindal 100, 118, 119, 131, 185, 221	250757
,, Cote 34, 82, 119, 131, 132, 135, 153, 238	
,, Moor 34, 78, 79, 82, 83, 84, 86, 148, 149, 230, 231, 236	247749

230, 231, 236	2577
,, valley 119	2476
Little Mill 35, 89, 95	219726
Liverpool 252	—
Liza, river 218	1613
Llantrissant 239	—
Lodore Falls 217	2618
Longsleddale 59, 180	4805
,, Chapel 68	5299
Lords Quarry 76	257943
Lorton Vale 219, 226	1526
Lowick (Berwick) 112, 114	—
,, (Furness) 75, 95	290860
,, High Common 100	2884
,, Low Common 100	2985
Loweswater 185, 195, 218, 219	1221
,, Church 198	142209
Lowther, river 216, 217	5124
Low Frith 252	340797
Low Hall Garth	338029
,, ,, (Ireleth) 29	233817
Low Mill 80	2977
Low Wood (Haverthwaite) 15, 16	345836
Low Wood Inn 18, 59, 180	386021
Lune, river 216	—
MANCHESTER 253	—
Marton 34, 77, 82, 149, 150	240771
Maryport 215	0437
Meathop Fell End 176	
,, Point 176	434792
Mere Tarn 115, 128,	267718
Millom 54, 59, 221, 250, 252	1780
,, Castle 25, 50	171815
,, Park 87, 88	1682
Millwood 36, 95	218730
Mint, river 216	5596
Mireside 219, 223	1015
Mirk Hall [Hole] 19	310020
Mississippi 116, 219	
Moat 93	278695
,, Hill 35	277697
Montmartre 129	
Moor Foot 35	207747
Moorhead 253, 254	233684
Moor Side 29, 72	226786
Morecambe Bay 9, 10, 13, 37, 87, 107, 108, 115, 128, 181, 217, 221, 251, 253, 254	
Moss Rigg 19	315025
Mountbarrow 109	2875
Mousell 33, 76, 152, 230	2375
Mull of Cantire [Kintyre] 133	—
Mwyndy (Glamorgan) 231, 232	—
NAB (Elterwater) 19	335042
Nan-beild, pass 217	4509
Nettleslack 24	230972
Newbarns 95	210706
Newbiggin 10, 35, 93	268693
Newby Bridge 16, 48, 71	369863
Newfield 24	228960
New Hall Estate 88	174795
Newland 15, 108	299797
Newlands 195, 206	237207
Newton 34, 89	230716
Nile, river 116, 219	—
North Wales 156	—
Nova Scotia 119	—

GAZETTEER

OLD HALLBECK [Holebeck] 100	234695
Old Hills 34, 77, 82, 176	244768
,, Park 252	3378
Orgrave 76, 142, 150	234760
,, Mill 76	233759
Ormsgill 95	195715
,, House [Hotel] 36	195715
,, Nook [Palace Nook] 37	188718
Osmotherly 158, 164	279821
,, Common 156	2681
Outcast 80	303777
Outerside 215, 222, 223, 226	115403
Outrake 71, 81, 85, 233	287796
Oxendale 20, 21, 23	2705
Oxford (Berwick) 112, 114	—
PALINGSBURN (Berwick) 112	—
Paris 129	—
Park 221	216754
Parkgate 69, 100	271843
Parkhead 13, 173, 174	334789
Park House 35, 89, 95	223711
Patterdale 216	3915
Pennington 85, 109	260774
Penny Bridge Hall 15	309833
Penrith 187, 213, 214, 216, 228	5231
Perm (Russia) 36	—
Petterill, river 216	4745
Piel Castle 10	233636
,, Channel 37	2364
Plumpton vii, 13, 14, 32, 74, 80, 106	313785
,, Hall 80, 94, 107, 108	313787
,, Wood 107, 252	312785
Point of Comfort 93	262681
,, ,, ,, Scar 93	263678
Powka Beck 142, 143, 150	2376
Prague 144	—
Preston 253	—
Pull Wyke 17, 58, 59, 180	365021
Purbeck 136	—
QUAY St. (Ulverston) 252	293785
Quarry Flat 170, 171	348768
RAISE GAP (Dunmail) 216-218	3211
Rakes Moor 95	207726
Rampside 10, 251, 253-258	2466
,, Church 254	239673
Raven Crag 182	334187
Ravenglass 217, 220	0896
Rawfold 24	200895
Rebecca Quarry 29	229783
Red Tarn 20, 21	267037
Red Pike 218	160154
Ribton 218	
Ricket Hills 33	230747
Roanhead 33, 89, 95, 221	203757
Rockcliff 216	3662
Rose Castle 216	371471
Roose 36, 95, 221	221696
,, Beck 10, 250 252, 255	258679
,, Cote 253	226689
,, Marsh 251	240685
Rosshead [Rosside] 75, 81, 82, 142-144, 147, 148, 157	271788
Rothay, river 18, 181, 217, 220	3605
Roudsea Wood 13, 16, 80, 174, 253	3382
Rumple Crag 19	
Rusland 71	3488
Rydal 181	3506
,, Mount 181	3606
SADDLEBACK 216	3228
St. Bees 166, 220	9415
,, Helen's viii, 34, 155	218746
,, John, Vale of 180, 181, 216, 217	3122
,, Lawrence 219	—
Saltcoats 217	079971
Sandscale 95	193739
Sandside 28, 70	226823
Sannibuts [Sawnbutts] 19	305015
Santon 217	1001
Scale Bank 77	236765
,, Hill 197, 226	154214
Scales 128, 221	270724
Scarth]hole 37, 39	183737
Scawfell 23, 185	207065
Scotland 53	
Scrimerstone (Berwick) 112, 114	—
Scrow 58	294977
Sea House [Sea Wood Farm] 112	290731
Seascale 220	037011
Seathwaite 20, 23-25, 43, 45, 91, 95, 217	2396
,, Parsonage 24	228961
Seaton Hall 96	107900
Sea Wood 11, 14, 93, 110	2973
Sebergham 216	358418
Selker 96	076887
Shap Wells 60, 68	579096
Shirestones 19, 23, 25	277027
Sinkfall 89, 95	212736
Skelwith Bridge 19	344035
,, Force 19	341035
Skiddaw 185, 211, 216, 227	2528
,, Forest 180, 193	2729
Skipton 129	
Snipe Gill 77, 82	244772
Solenhopen 52	—
Solway, Firth of 215, 216	
Soutergate 28, 70, 76	227814
Sowerby Hall 36, 95	199725
,, Lodge 95	191723
Stainton 11, 34, 83, 92, 95, 107, 114, 135-138, 140, 221, 231, 237, 252	248725
,, Head 138	243724
Stake Pass 217	2608
Standing Stones 88	133809
Stang End 19	319029
Stank 91, 92, 97, 100	234705
Stewnor Bank 30	236781
Stone Close 85	2573
,, Dykes 69, 100	273852
Stonster 24	201912
Styhead Pass 217, 219	2109
Sty Rigg 19	312024
Sunderland Terr. (Ulverston) 252	293785
Sunnybank 67	290925
Sunny Brow 58	342004
Swarthmoor 80, 81, 114, 221	274773
,, Hall 80, 81	282774

Swarthdale House 80, 81, 100, 109	
	286773
Sylecroft 87, 220	130820
TARN 96	078898
,, Close 77, 81, 82, 143, 150	267785
Terry [Rectory] 93	286711
Thirlmere 181	3116
Thornbarrow 216	4097
Threlkeld 216	3125
Throng 24	236977
Tilberthwaite 19, 43, 45	305011
Tippin's Bridge 29, 72	226793
Todhillbank Farm [Tottlebank] 68, 70, 71	270882
,, Fell 68, 73	2688
Torpenhow 215	2039
Torver 61, 67, 99	285942
,, Chapel 68	285943
,, Common 53, 62	2896
Town Beck (Coniston) 58	302977
,, ,, (Ulverston) 142	287787
Tredlea Point [Tridley] 14	323792
Trinklet [Trinkeld] 115	276765
Troutbeck 26, 59, 180	410035
,, , river 217	4204
Trouthall 24	235988
Tunnel 19	313020
Turner Hall 24	233965
ULLSWATER 216, 217	4220
Ulpha 43, 44, 217	1993
Ulverston 75, 82, 85, 95-97, 99, 100, 108, 115, 116, 131, 136, 142, 149, 156, 179, 194, 220, 228, 229, 233, 252	2878
,, Canal 106	3078
,, Sands 9, 15, 80, 250	3174
,, Station 109	285779
Undercrag 24	233967
Underhill 61	181830
Ure Pits 33, 82	240753
Urswick 95, 116, 221	2673
,, Green iii, vi, 123, 124	263737
,, , Little 123	263736
,, , ,, , Tarn 122	266739
,, , Much [Great] 115, 118, 119, 120	269746
,, Tarn 117, 121, 122, 128, 131, 132	270745
VICKERS 19	298030

WADHEAD 93	310745
,, Scar vii, 12	309745
Waitham Hill 66, 70	224846
Wales 98, 176	
Wall-end (Foxfield) 27	221879
,, ,, (Kirkby) 28, 66, 67	235831
,, ,, (Langdale) 2	283055
Walls Moss 67	
Walney forest 252	186653
,, Island 9, 19, 25, 26, 37-39, 40, 42, 250, 251	1867
,, Lighthouse 10	230620
,, Scar [Walna] 185	257965
,, ,, Quarry 43-45	248960
Wastdale 23	1606
,, Head 217	187088
Wastwater 217	1606
Waterblain 25, 50	177825
Waterend 195	1122
Waterfoot 71	291897
Waterhead 55, 63, 64, 68, 74	318983
Wateryeat 67	289891
Waver, river 216	2545
Waverton 216	2347
Wegbarrow Point 24	234995
Well Wood 11, 110	297742
Westfield Point 255, 256, 258	224668
Westmorland 10, 19, 48, 69, 98, 130	
Whampool, river 216	2555
Whicham Valley 87	1584
Whineray Ground 24	201904
Whinfield 34, 78, 80	252765
,, Point 82, 85	257764
Whistleton Green 24	195927
Whitehaven 185-187, 220, 223	9718
Whiteside 196, 197, 226, 228	170220
Whitriggs 34, 82	242761
,, Bottom 34	245760
Wigton 216	2548
Windermere 9, 10, 15-19, 25, 48, 68, 91, 99	3995
,, Ferry 16, 17	390956
,, (lake) 59, 61, 63, 64, 71, 181, 217	3995
Winder Moor 252	3775
Winster, river 176	4286
Woodland 61, 67, 70	241903
Workington 182, 213, 220, 223	0029
Wray 216	3700
Wreaks Causeway 99	2286
,, -end 27	223866
,, Moss 174	3482
Wrynose 19, 23, 45, 217	2702
Wythburn 180-182, 217	324137
YARL WELL 150	235743

DRAMATIS PERSONAE

Abbot of Furness (Robert de Denton) 128
Aesop 56
Ainslie, W. 79
Ainsworth, T. 16
Aldingham, Lords of 128
Archibald, C. D. 42, 119
Ashburner, Geo. 158
Ashburner, Rev. William 4, 123
Ashburner & Sons 33
Atkinson, Mr. 184
Atkinson, D. Frontispiece
Atkisson, Seppy 147
Aveline, W. Talbot 174

Balfour, Dr. 132
Barber, Dr. 170
Barker, Misses 16
Barrande, M. 144
Barret, Rev. W. C. 104
Barrett, Mr. 46, 58
Barrow Haematite Iron & Steel and Mining Co. (see Schneider, Hannay & Co.) 34, 152, 221
Beccaria 258
Binney, E. W. 1, 148, 149, 243
Bolton, John 131, 166, 210
Boulton, W. 89
Bowman, Stephen 212
Braddyll, T. R. G. 11, 138
Brogden, A. M.P. Inscription, 6, 8, 83, 84, 105, 109, 152, 153, 175, 231, 232, 234
Brogden, A. & Co. 231
Brogden, John 6, 14
Buccleuch, Duke of 78, 90

Cameron, A. C. Grant 174, 238
Casarus, Owen 88
Clegg, J. 34, 254, 255
Clegg & Co. 155
Coleridge, Hartley 181
Coupland, Sir Richard de 128
Coward, Edward 29

Cowper 105
Cranke, Malachi 2, 124, 128
Cross, R. Assheton, M.P. 66
Cuvier, Baron 129

Denney, James 155
Denney & Co. 34, 155
Des Cartes 258
Devonshire, Duke of 14, 36, 116, 128, 158, 160, 163, 172 - 174, 252
Dick, A. 235, 236, 241
Drewry, G. 172
Dunmail, King of Cumberland 181

Edmund - Saxon Monarch 181
Evans, Rev. Francis 88, 234

Fearn, Isaac 223
Fell, Margaret 114
Fell, J. 234
Fox, George 114

Gale, Capt. R. H. 12
Garden, J. 11, 136
Glass, Mrs. 225
Gordon, Jean 226
Gough, Dr. 29, 202
Graham, Joseph 194, 213
Graindorge, Sir George 137
Gwillym, Rev. Canon 104, 148

Harkness, Prof. 36, 178, 179, 182, 183, 190, 194, 197, 202, 212, 224, 225
Harrison, Ainslie & Co. 15, 34, 78, 108
Hasting, Mary 186 - 193
Hier, Prof. 135
Hodgson, Miss E. 132
Huddleston, G. 135
Huddleston, Messrs. T. & W. 253

Jackson, John 206
James, Cousin 5
James IV, King 112

Jopling, C. M.   46, 155, 160

Kennedy, Mrs.   81
Kennedy Bros., Messrs.   30, 33, 221
King of England   94
King, Prof.   1, 202
Kircher   258
Kirkby, George   116

Machell, J. P.   15
Mallet, Mrs.   258
Marshall, James Garth   55, 63, 68
M'Coy, Prof.   244
Middleton, John O.   166
Miller, Prof. Hugh   5, 53, 97, 242
M'Noad, Dr. H.   239, 241
Moon, A.   229 - 231
Moore, Thomas   15
Morris, J. P.   168, 182
Mucklow, E.   175
Muncaster, Lord   78
Murchison, Sir Roderick   36

Napoleon I   142
Nicholson, Henry Alliyne   222, 224, 225

Oldbuck, Johnathan   123
Ormandy, Mr.   128

Pearson, Henry   195
Petty, Mrs.   11, 110
Phillips, Prof.   1, 82, 149, 202
Playfair, Dr. Lyon   237
Postlethwaite, Miss   122
Postlethwaite, Mr.   62
Postlethwaite, Messrs.   24, 163
Priestley, Dr.   258

Ramsden, J.   35
Rawlinson, J.   33, 34, 77, 80, 83, 150, 151
Rawlinson, W. S. Major   25
Rawlinson & Briggs   81
Rawlinson & Shaw, Messrs   32
Rigg, John   206
Roberts, G. E.   166
Roberts, G. S.   210

Rogers, Prof.   257, 258
Ross, Mr.   253

Salmon, W.   166, 234
Salt, S.   243
Salter, J. W.   6, 7, 75, 81, 184, 210
Sawrey, John   62
Schneider, Hannay & Co.   33, 34, 37 (see Barrow Haematite I. S. M. Co.)
Scott, Sir Walter   113, 154, 181
Sedgewick, Prof. Rev. Adam   1, 2, 23 - 25, 28, 29, 32, 48, 54, 57, 59, 64, 65, 68 - 71, 73 - 75, 112, 179, 194, 202
Simpson, William Parker   255 - 257
Sorby, H. C.   230, 231
Soyer, M.   225
Spiller, J.   236
Stephen, Mr.   132
Stephen, Earl of Moreton & Bologne   94
Stukely, Dr.   258
Sunderland, George   80, 100
Surrey, Earl of   112
Swainson, W. T.   243

Thompson, John   255, 257
Thorpe, E. T.   170
Tolming, Rev. T.   58, 122

Ulverston Mining Co.   33, 34, 131

Vivian   232

Wadham, E.   34, 35, 89, 90
Wales, Princess of   226
Walker, Rev. Robert   24
Wallich, Dr.   133
Walton, Isaac   145
Webb, Maria   114
Wilkinson, J.   175
Wilson, Prof.   226
Woodward, Mr.   7
Wordsworth, William   24, 32, 48, 69, 73, 75, 181
Wright, Charles   184

Young, J. S.   166